TEACHING AND LEARNING MATHEMATICS IN CONTEXT

T. BREITEIG,
Kristiansand Laererhogskole, Norway
I. HUNTLEY,
Sheffield Hallam University
G. KAISER-MESSMER,
Department of Mathematics, Kassel University, Germany

ELLIS HORWOOD
NEW YORK LONDON TORONTO SYDNEY TOKYO SINGAPORE

First published 1993 by
Ellis Horwood Limited
Market Cross House, Cooper Street
Chichester
West Sussex, PO19 1EB
A division of
Simon & Schuster International Group

Transferred to digital print on demand 2002

Printed and bound by Antony Rowe Ltd, Eastbourne

Library of Congress Cataloging-in-Publication Data

Available from the publisher

British Library Cataloguing in Publication Data

A catalogue record for this book is available from the British Library

ISBN 0-13-031006-9 (hbk)

1 2 3 4 5 97 96 95 94 93

TEACHING AND LEARNING MATHEMATICS IN CONTEXT

MATHEMATICS AND ITS APPLICATIONS

Editor: B.W. CONOLLY

Emeritus Professor of Mathematics (Operational Research), Queen Mary College, University of London

Anderson, I	**COMBINATORIAL DESIGNS Construction Methods**
Arczewski, K. & Pietrucha, J.	**MATHEMATICAL MODELLING OF COMPLEX MECHANICAL SYSTEMS: Volume 1: Discrete Models**
Arczewski, K. & Pietrucha, J.	**MATHEMATICAL MODELLING OF COMPLEX MECHANICAL SYSTEMS: Volume 2: Continuous Models**
Baker, A.C. & Porteous, H. L	**LINEAR ALGEBRA AND DIFFERENTIAL EQUATIONS**
Balcerzyk, S. & Josefiak, T.	**COMMUTATIVE RINGS**
Barnett, S.	**SOME MODERN APPLICATIONS OF MATHEMATICS**
Beaumont, G.P. & Knowles, J.D.	**STATISTICAL TESTS: An Introduction with Minitab Commentary**
Bejancu, A.	**FINSLER GEOMETRY AND APPLICATIONS**
Bell, G. M. & Lavis, D.A.	**STATISTICAL MECHANICS OF LATTICE MODELS, Vols. 1 & 2**
Breiteig, T., Huntley, I. & Kaiser-Messmer, G.	**TEACHING AND LEARNING MATHEMATICS IN CONTEXT**
Bunday , B.D.	**STATISTICAL METHODS IN RELIABILITY THEORY AND PRACTICE**
Berry, J.S., Burghes, D.N., Huntley, I.D., Bunday , B.D.	**STATISTICAL METHODS IN RELIABILITY THEORY AND PRACTICE**
James, D.J.G. & Moscardini, A.O.	**MATHEMATICAL MODELLING COURSES**
Berry, J.S., Burghes, D.N., Huntley, I.D., James, D.J.G. & Moscardini, A.O.	**MATHEMATICAL MODELLING METHODOLOGY, MODELS AND MICROS**
Berry, J.S., Burghes, D.N., Huntley, I.D., James, D.J.G. & Moscardini, A.O	**TEACHING AND APPLYING MATHEMATICAL MODELLING**
Berry, J.S., Graham, E. & Watkins, A.J.P.	**LEARNING MATHEMATICS THROUGH DERIVE**
Burghes, D.N. & Borrie, M.	**MODELLING WITH DIFFERENTIAL EQUATIONS**
Beaumont, G.P. & Knowles, J.D.	**STATISTICAL TESTS: An Introduction with Minitab Commentary**
Cartwright, M.	**FOURIER METHODS: for Mathematicians, Scientists and Engineers**
Cerny, I.	**COMPLEX DOMAIN ANALYSIS**
Chorlton, F.	**VECTOR AND TENSOR METHODS**
Cohen, D.E.	**COMPUTABILITY AND LOGIC**
Crapper, G.D.	**INTRODUCTION TO WATER WAVES**
Cullen, M.R.	**LINEAR MODELS IN BIOLOGY**
Doucet , P.G. & Sloep, P.B.	**MATHEMATICAL MODELING IN THE LIFE SCIENCES**
Dunning-Davies, J.	**MATHEMATICAL METHODS FOR MATHEMATICIANS, PHYSICAL SCIENTISTS AND ENGINEERS**
Eason, G. Coles, C.W. & Gettinby, G.	**MATHEMATICS AND STATISTICS FOR THE BIOSCIENCES**
Farrashkhalvat, M & Miles, J.P.	**TENSOR METHODS FOR ENGINEERS AND SCIENTISTS**
Faux, I.D. & Pratt, M.J.	**COMPUTATIONAL GEOMETRY FOR DESIGN AND MANUFACTURE**
French, S.	**SEQUENCING AND SCHEDULING: Mathematics of the Job Shop**
French, S.	**DECISION THEORY; An Introduction to the Mathematics of Rationality**
Firby, P.A. & Gardiner, C.F.	**SURFACE TOPOLOGY: Second Edition**
Gardiner, C.F.	**ALGEBRAIC STRUCTURES**
Gilbert, R.P. & Howard, H.C.	**ORDINARY AND PARTIAL DIFFERENTIAL EQUATIONS WITH APPLICATIONS**
Goodbody, A. M.	**CARTESIAN TENSORS**
Griffel, D.H	**APPLIED FUNCTIONAL ANALYSIS**
Griffel, D.	**LINEAR ALGEBRA AND ITS APPLICATIONS: Vol. 1, A First Course; Vol. 2, More Advanced**
Griffiths, P. & Hill, I.D.	**APPLIED STATISTICS ALGORITHMS**
Griffiths, H.B. & Oldknow, A.	**MATHEMATICAL MODELS OF CONTINUOUS AND DISCRETE DYNAMIC SYSTEMS**
Guest, P.B.	**LAPLACE TRANSFORMS AND AN INTRODUCTION TO DISTRIBUTIONS**
Hart, D. & Croft, A.	**MODELLING WITH PROJECTILES**
Hartley, R.	**LINEAR AND NON-LINEAR PROGRAMMING**
Hoskins, R.F.	**STANDARD AND NONSTANDARD ANALYSIS**
Hoskins, R.F. & de Sousa Pinto, J.J.M.	**DISTRIBUTIONS, ULTRADISTRIBUTIONS AND OTHER GENERALIZED FUNCTIONS**
Irons, B.M. & Shrive, N.G.	**NUMERICAL METHODS IN ENGINEERING AND APPLIED SCIENCE**
Janacek, G. & Swift,	**TIME SERIES: Forecasting, Simulation, Applications**
Johnson, R.M.	**THEORY AND APPLICATIONS OF LINEAR DIFFERENTIAL AND DIFFERENCE EQUATIONS**
Johnson, R	**CALCULUS: Theory and Applications in Technology and the Physical and Life Sciences**

series continued at the back of the book

Contents

2.4 Tertiary education

Section 3
LOOKING AHEAD –
RESEARCH ON MATHEMATICAL MODELLING

Preface

I am convinced that, if we do not succeed in teaching mathematics so as to be useful, users of mathematics will decide that mathematics is too important a teaching matter to be taught by the mathematics teacher. Of course this would be the end of all mathematical education.
(H Freudenthal, 1968)

The history of mathematics demonstrates that the subject has always been closely related to other scientific areas, to philosophy, and to life in all its different aspects. This mutual impact has stimulated the development of mathematics, and also the development of other areas – we find this interaction in the time of Kepler, with people looking for universal laws to explain the solar system, and we see it in modern studies in information theory and network.

The relationship between mathematics and other areas of reality and life has been a core trend in mathematics education during the last decades. These aspects are also emphasised in mathematics curricula – mathematics as a servant of life, and dependent on the problems posed in other areas. The modelling of a complex situation from daily life or from other subjects can help to overview the situation, make a decision, make a prognosis – this gives the *raison d'être* for mathematics for many individuals. Students can see, in small glimpses, the potential lying in the dialogue between mathematics and other areas of reality. However, many questions do arise when introducing these aspects into teaching, and we discuss some of these below.

What is the relationship between modelling and applications, on the one hand, and the development of the necessary mathematical tools on the other? How can students decide upon modelling when they don't have the mathematical tool necessary – and how can the pure mathematical tools be motivated without applications? How is it possible to give authentic, non-trivial models and applications which are of genuine interest within the frames of education? The development of which materials is most urgently needed? How can we utilise and share what is already developed, and the experience gained through investigations?

This trend in mathematical education was evident at the Lausanne Symposium in 1968, which has become famous under its theme *How to teach mathematics so as to be useful.* The symposium pointed to the utilisation of mathematical structures in reality as a major objective of mathematics teaching. This does not mean the teaching of closed applications, but the ability to mathematise and model extra-mathematical problems and situations. Further reformation of mathematics teaching has to be considered as a permanent process, and implies a continuous retraining of teachers based on ongoing pedagogical research in a global collaboration – mathematical modelling is a continuing stream in mathematical education. This book has its place in this stream.

The book is a collection of papers, many of which were initially presented at Working Group 14, *Mathematical modelling in the classroom,* at the seventh International Congress on Mathematical Education, ICME-7, in Quebec (16-23 August 1992). Chief organiser for this group was Trygve Breiteig, and the advisory team was Gabriele Kaiser-Messmer, Jan de Lange, and Zalman Usiskin, with Brendan Kelly as the local organiser. The group was in a part of the congress subdivided according to the level of education – primary, lower-secondary, upper-secondary and tertiary education – guided by Leen Streefland/Heleen Verhage, Alan Rogerson, David Burghes/Gabriele Kaiser-Messmer and John Berry respectively. An outline of the reports and discussions can be found in the main congress proceedings.

A state-of-the-art overview, and an outline of the theoretical base for the Working Group, was given in the lecture by Werner Blum, which forms the basics for Chapter 1. The following chapters covers different issues – they contain examples, ideas of teaching materials developed and experience gained, as well as empirical results and theoretical reflections. In the last session of the Working Group, the members looked ahead with a research perspective – this is emphasised in the chapters by Solomon Garfunkel, Gabriele Kaiser-Messmer, and Zalman Usisikin, which are included in the last section of the book.

Thus the issues concerning the relations between mathematics and other areas of the universe, and how these are being implemented in the curriculum, are continued from the previous International Congresses on Mathematical Education (the ICMEs), and the International Conferences on the Teaching of Mathematical Modelling and Applications (the ICTMAs). Of course, this area is a very wide area – both in educational range and from the mathematical point of view. This book will therefore have a wide readership – from those interested in students' primary mathematical work with numbers and space, to those concerned with education in technology and engineering, and those carrying out research into modelling.

We thank all the contributors to this book. Thanks also to our own institutions for the support during the work on the book, and to the publishers, Ellis Horwood, for making it possible for this work to be shared more widely.

Kristiansand/Sheffield/Kassel June 1993

Trygve Breiteig *Ian Huntley* *Gabriele Kaiser-Messmer*

Section 1

What is mathematical modelling?

1

Mathematical modelling in mathematics education and instruction

Werner Blum
Mathematics Department, Kassel University, Germany

SUMMARY

This paper aims at giving a concise survey of the present state-of-the-art of mathematical modelling in mathematics education and instruction. It will consist of four parts. In part 1, some basic concepts relevant to the topic will be clarified and, in particular, mathematical modelling will be defined in a broad, comprehensive sense. Part 2 will review arguments for the inclusion of modelling in mathematics teaching at schools and universities, and identify certain schools of thought within mathematics education. Part 3 will describe the rôle of modelling in present mathematics curricula and in everyday teaching practice. Some obstacles for mathematical modelling in the classroom will be analysed, as well as the opportunities and risks of computer usage. In part 4, selected materials and resources for teaching mathematical modelling, developed in the last few years in America, Australia and Europe, will be presented. The examples will demonstrate many promising directions of development.

1 WHAT IS MATHEMATICAL MODELLING?

There are as many definitions of mathematical modelling as there are authors writing about it – quite apart from the spelling with one 'l' or two – and the annual number of publications on modelling is increasing steadily. Let me mention some useful recent books in English that deal with the question of what modelling is: Blum, Niss and Huntley (1989), Clements (1989), Edwards and

Hamson (1989), Huntley and James (1990), Murthy, Page and Rodin (1990), Starfield, Smith and Bleloch (1990). Of course, older books such as Cross and Moscardini (1985) or Giordano and Weir (1985) are still useful, too, as well as the series of proceedings of the International Conferences on the Teaching of Modelling and Application (ICTMA): Berry *et al* (1984), Berry *et al* (1986, 1987), Blum *et al* (1989), Niss, Blum and Huntley (1991), de Lange *et al* (1993).

Now, as a basis for the following parts, I shall give some pragmatic working definitions which have been widely accepted in mathematics eduction in recent years (see the survey article by Blum and Niss, 1991). Let me quote the well-known simple model of *applied mathematical problem solving*.

The starting point is a *situation* in the real world, that means in the rest of the world outside mathematics. The situation normally has to be simplified, structured and made more precise by the problem solver, which leads to a *real model* of the situation. The real model is not – as one will often read – merely a simplified but true image of some part of an objective, pre-existing reality. Rather the step from situation to model also creates a piece of reality, dependent on the intentions and interests of the problem solver.

Then, if possible (if there is some mathematics in it), the real model is *mathematised*, that is translated into mathematics, resulting in a *mathematical model* of the original situation. Sometimes, different models of the same situation may be constructed.

The problem-solving process continues by choosing suitable methods and working within mathematics, through which certain *mathematical results* are obtained. These have to be retranslated into the real world – that is, to be *interpreted* in relation to the original situation. In doing so, the problem solver also validates the mathematical model. If discrepancies occur – which will often happen in reality, since there are so many potential pitfalls (see Murthy, Page and Rodin, 1990) – then the whole cycle has to start again.

Actually, all of the above is valid only for 'really real' situations. Sometimes – especially in school mathematics – the given situation is just an artificial dressing-up of some purely mathematical problem. Then the model-building process consists merely of an undressing of this *word problem*, and the problem-solving process stops after only one cycle. Nevertheless, such artificial problems may be of didactical value.

The term *application* of mathematics may be used for different parts of this cycle. A real-world situation may be called an application, using mathematics to investigate real situations may be denoted applying mathematics, or any manner of connecting the real world with mathematics can be seen as an application of mathematics. Equally, the term mathematical *modelling* may mean the process of model building, leading from a real situation to a mathematical model, or the whole applied problem-solving process, or again any manner of connecting the real world with mathematics. In recent years, the term *applications and modelling* (or vice versa) is frequently used as an all-embracing expression for the various interrelations just mentioned. In the following, in order to simplify matters, I will use the term (mathematical) *modelling* mostly in this comprehensive and extensive sense. This terminology is also justified by the fact that there is an international trend towards broadening the views on applications, models, mathematising, modelling, links between mathematics and other subjects, and so on. Thus the relations between these components have become stronger in the last ten years, both in research and in practice, as identified in Blum and Niss (1991).

2 WHAT IS MODELLING GOOD FOR IN MATHEMATICS TEACHING?

In the last 15 years there has been a worldwide trend towards modelling (in the broad sense) in mathematics education, at school and university levels. In particular, there has been a recent shift from applications to modelling, that is a trend towards emphasising the translation processes and not only working with ready-made models.

Today, modelling is favoured by almost everybody, as are problem solving, active learning and other activities of this nature (see, for example, the Cockcroft Report in the UK or the NCTM Recommendations in the USA). There are various *reasons* for favouring modelling in mathematics teaching. I would like to mention briefly four essential arguments, based mainly on general goals and aims for mathematics instruction (see Blum, 1991).

- *Pragmatic arguments.* Mathematics teaching is intended to help students to understand and to cope with real-world situations and problems. To that end, modelling is indispensable.
- *Formative arguments.* By being concerned with mathematics, students should – we hope – acquire general qualifications (such as the ability to tackle problems) or attitudes (such as openness towards new situations).

Modelling is one important way to develop these.

- *Cultural arguments*. Students should be taught mathematical topics as a source for reflection, or in order to generate a comprehensive and balanced picture of mathematics as a science and a part of human history and culture. Modelling is an essential feature of human intellectualism as well as of history and of actual practice, and can thus contribute towards promoting those aspects.
- *Psychological arguments*. Mathematical contents can be motivated or consolidated by suitable modelling examples, and these may contribute towards deeper understanding and longer retention of mathematical topics, or they may improve students' attitudes toward mathematics.

Let me put it even more generally. At all times and places there have been complaints about the lack of sense and the absence of meaning in mathematics learning and teaching. Students very often experience mathematics as a mechanical manipulation of meaningless symbols. Modelling can contribute towards giving more *meaning* to the learning and teaching of mathematics. If the formation of appropriate basic ideas and the development of meaning are essential aims of mathematics instruction, then modelling has to be an integral part.

The debate in mathematics education on modelling is not uniform. Different persons and groups emphasise different aims for mathematics teaching, and different arguments for the inclusion of modelling. Accordingly, it is possible to identify certain *schools of thought* within mathematics education. Kaiser-Messmer (1991) has discerned three of them:

- a so-called *pragmatic* school, broadly accepted in the English language area, where pragmatic aims are placed in the foreground;
- a so-called *scientific-humanistic* school, widely accepted in the Romance language area, where cultural aims are stressed;
- a so-called *integrated* school, largely found in the German language area.

However, since the eighties, there has been a trend in mathematics education all over the world towards widening the spectrum of aims and arguments, and thus to accept all arguments as important. In any case, no further advertising seems to be necessary, as there is a consensus of opinion in the mathematics education community that modelling has to be an integral part of mathematics instruction. Now, what is the actual situation in mathematics curricula and especially in school and university classrooms? Here, in my experience, a very different picture arises.

3 WHAT IS THE ROLE OF MATHEMATICAL MODELLING IN CURRICULA AND IN EVERYDAY TEACHING PRACTICE?

When considering modelling in mathematics *curricula*, there are differences with respect to

- the aims of the mathematics teaching and the corresponding rôle of modelling,
- the range of mathematical topics with modelling content,
- the areas that modelling examples are taken from (science, economy, everyday life, sports and so on),
- the kinds of examples (global – local, authentic – artificial, open – closed, relevant – irrelevant for students, and suchlike)
- the approaches to the organisation of mathematical and modelling curriculum components (separated, mixed, integrated).

Globally speaking, there is a clear world-wide trend towards including more modelling into mathematics curricula, also – though to a smaller extent – modelling in the strict sense. Some present-day secondary school curricula, for instance from the Netherlands or from Victoria in Australia, have included compulsory modelling components throughout. Furthermore, both the range of applied mathematical topics and of applicational areas have broadened – for example, more so-called discrete mathematics is included, at all levels from lower secondary to tertiary.

In everyday *teaching practice*, the amount of mathematical modelling work is also increasing. Particularly standard models – small, well-defined examples which are easy to survey and fit in well with the conventional curriculum – have entered the classroom. However, there is still a substantial gap between the forefront of research and development in mathematics education, on the one hand, and the mainstream of mathematics instruction, on the other hand. In most countries, modelling (in the broad and, even more so, in the strict sense) still plays only a minor rôle in everyday teaching practice at school and university. Formal calculations prevail, and applicational examples are treated, if at all, only as illustrations, and are not taken seriously as such. Even competent and committed teachers often do not follow all those well-founded recommendations and suggestions of mathematics educators. Why is this?

I think it is not due to ill-will or incompetence in teachers but to actual barriers and obstacles. Let me summarise briefly three of these.

- Obstacles from the point of view of *instruction and assessment*. Modelling requires a lot of time, and fits neither the regular mathematics syllabus (which is overloaded anyway) nor the regular school or university organisation (with 45 minute lessons and suchlike). Modelling is difficult to assess (see Niss, 1993), and what is not examined will not be taken seriously by students or by teachers.
- Obstacles from the *student's* point of view. Modelling makes mathematics lessons and examinations more demanding and less predictable. We have renewed evidence of this fact from our recent comparative empirical studies in England and Germany (see Blum *et al,* 1992).
- Obstacles from the *teacher's* point of view. Modelling makes teaching more demanding, because additional non-mathematical knowledge and qualifications are required, as well as the ability to manage open situation in the classroom. So the teacher's rôle is changed. Furthermore, teachers often do not know enough modelling examples suitable for instruction, and the ones they have at hand have to be brought up-to-date and have to be adapted for the class, for which a great deal of additional effort has to be invested.

One might think that the last point is a fourth obstacle, namely that there are not enough modelling examples and materials suitable for teaching. I am convinced that this is not true. For nearly every mathematical topic area there is a wealth of modelling examples of all kinds, in textbooks, case collections, teaching units, books or articles, all in principle accessible to teachers. I will name a few recent materials below.

There is a lot being done to overcome those obstacles, in curriculum development or in pre-service and in-service teacher education. According to my observations, the situation in the classroom is in the process of being improved. In countries all over the world, teachers in schools, colleges and universities are becoming more and more aware of deficiencies and trying to bring about change in the desired directions – towards a broader range of aims, towards more applications, towards more authentic examples, towards more open problems, and so on. Therefore there is a clear trend towards reducing the gap mentioned earlier, though with considerable differences between various countries.

I am convinced, though, that the barriers for modelling are very difficult to surmount. I see two things especially that need to be done. First, to initiate all kinds of teacher education activities, in order to supply as many teachers as possible with knowledge, abilities and, in particular, with attitudes to cope with the demands of teaching modelling – up to now, only very few reports on modelling experiences in teacher education exist. Second, to develop appropriate modes of assessment. Here we can learn a lot from the Dutch and the Australian materials and from their experiences, both good and bad. All these kinds of activities have to be accomplished essentially by teachers themselves.

There is some hope that the situation in mathematics classroom will improve especially through the increasing availability of powerful *computers*. Without doubt, new exciting opportunities have emerged for making contents accessible to learners, or for relieving mathematics teaching and learning of some tedious activities. This has obvious implications for mathematical modelling, too. In fact, there is a trend towards an extended use of computers in mathematics teaching. However, we should not expect too much. First, the obstacles just mentioned remain largely unaffected by the presence of computers; on the contrary, even more qualifications and higher-level abilities are demanded from students and teachers. Second, computers may entail many kinds of new problems and risks – for instance, students may try to replace necessary intellectual effort with modelling by mere button pressing. Third, there are and will be organisational problems with the accessibility of personal computers. I am convinced that nothing but powerful and cheap pocket computers will really affect mathematics teaching.

4 WHAT MATERIALS ARE AVAILABLE FOR TEACHING MATHEMATICAL MODELLING?

As to older resources, I refer to the surveys from the International Congresses on Mathematical Education (ICME): Pollak (1979), Bell (1983), Niss (1987), Blum and Niss (1991).

In the last few years, many interesting new materials have been developed. Let me present a small, subjective selection, concentrating on the secondary-school level.

From Australia

- Two booklets by Carr and Galbraith (1987, 1991) from the project PAM, with detailed examples and teaching units for the lower-secondary level, covering a broad range of extra- and intra-mathematical subjects.
- Two volumes by Lovitt and Clarke (1988), an extensive collection of detailed examples for the lower-secondary level, oriented towards students' activities.

- Two books by Lowe (1988, 1991), with numerous detailed examples for grades 7–12, covering a broad range of subjects, mostly referring to computers as a tool.
- In Australia, there are also many efforts towards new assessment modes for modelling-oriented curricula (see, for example, Clatworthy and Galbraith, 1991).

From the Netherlands

- The Freudenthal Institute (before 1991 known as OW&OC) at Utrecht University has developed lots of materials, including textbooks, for all grades in school: primary (the Wiskobas project by Treffers *et al*), upper-secondary (the Hewet project by de Lange *et al)*, and more recently also lower-secondary. Many materials have also been published in English, especially in connection with the Madison project (see de Lange, 1992). All materials are structured according to mathematical topics, and the modelling examples are intended to support the learning of the mathematics.

From the UK

- Several booklets and materials from the Numeracy Through Problem Solving project, developed in the Shell Centre for Mathematical Education at Nottingham University (involving Burkhardt *et al*).
- The Enterprising Mathematics Course from the Centre for Innovation in Mathematics Teaching at Exeter University (by Burghes *et al*).
  Both the Shell and the CIMT materials are written for the lower-secondary level, and both are structured according to problems, not to mathematics, and aim at stimulating students' activities.
- A series of booklets for the A-level from the Spode Group (1992), with detailed examples for direct use in the classroom.
- I also refer once more to the books mentioned at the beginning of this paper, all containing many case studies, mostly for the tertiary level.

From the USA

- Two series of books with numerous modelling examples from the Consortium for Mathematics and its Applications (involving Garfunkel, Aragon, Malkevitch *et al*) are the HIMAP Modules (1985-92) for the secondary level and the UMAP Modules (1981-92) for tertiary and upper-secondary levels.
- The book by Garfunkel and Steen (1991), an introduction to recent mathematical applications in practice, especially in connection with computers, suitable for secondary and tertiary levels, with some mathematical topics as yet outside present syllabi.

- A series of application-oriented textbooks (1989-92) from the University of Chicago School Mathematics project (involving Usiskin, Bell *et al*), covering arithmetic, algebra, geometry, statistics, precalculus and discrete mathematics.
- From NCTM an introduction to college mathematics (1988), developed by the North Carolina School of Science and Mathematics (Teague *et al*), and a booklet by Swetz and Hartzler (1991) with modelling examples for the secondary level, oriented towards students' activities.

From Germany

- The project MUED (Mathematik-Unterrichtseinheiten-Datei involving Böer, Volk *et al*) has developed several detailed global teaching units aiming at students' abilities to act competently in real-life situations. Two examples for the lower-secondary level are Böer (1990) and Jannack (1992).
- Several projects at the university level deal with real uses of mathematics in industry, and their utilisation in the training of future mathematicians or mathematics teachers. An example is the case studies in Knauer (1992).

Many more references to literature for the secondary-school level in German, English and French, are contained in the extensive bibliography by Kaiser-Messmer, Blum and Schober (1982/1992).

Altogether, these examples demonstrate many promising *directions of development* from various parts of the world.

- Application-oriented mathematics curricula on a broad scale in the Netherlands.
- Case collections for new modelling-oriented mathematics curricula, including assessment, in Australia.
- Examples of activity-oriented global-problem sequences in England.
- Materials for a new orientation of mathematics curricula in the computer age in the USA.
- Application-oriented teaching units as alternatives to conventional mathematics instruction in Germany.

In addition to these, there are many extremely interesting activities in several other countries that I could not present in this paper. Of course, many past and present noteworthy activities in the field of learning and teaching mathematical modelling are unknown to the public simply because of language problems or since only non-official publications exist. However that may be, I apologise for any omissions.

So, looking ahead, in my view we could and should be rather optimistic, despite a number of still unsolved or perhaps even unsolvable problems. I trust that this will not only be my view – optimism is probably a necessary universal attitude for mathematics teachers and educators.

REFERENCES

Bell, M (1983). Materials available worldwide for teaching applications of mathematics at the school level. In Zweng, M *et al* (ed), *Proceedings of the Fourth International Congress on Mathematical Education.* Boston: Birkhäuser, 252-267.

Berry, J *et al* (ed) (1984). *Teaching and Applying Mathematical Modelling.* Chichester: Ellis Horwood.

Berry, J *et al* (ed) (1986). *Mathematical Modelling Methodology, Models and Micros.* Chichester: Ellis Horwood.

Berry, J *et al* (ed) (1987). *Mathematical Modelling Courses.* Chichester: Ellis Horwood.

Blum, W (1991). Applications and modelling in mathematics teaching - a review of arguments and instructional aspects. In *Niss, M, Blum, W and Huntley, I* (ed), loc. cit., 10-29.

Blum, W *et al* (ed) (1989). *Applications and Modelling in Learning and Teaching Mathematics.* Chichester: Ellis Horwood.

Blum, W *et al* (1992). Teaching and learning of mathematics and its applications: first results from a comparative empirical study in England and Germany. In *Teaching Mathematics and its Applications,* **11**, 112-123.

Blum, W and Niss, M (1991). Applied mathematical problem solving, modelling, applications, and links to other subjects - state, trends and issues in mathematics instruction. In *Educational Studies in Mathematics,* **22**, 37-68.

Blum, W, Niss, M and Huntley, I (ed) (1989). *Modelling, Applications and Applied Problem Solving - Teaching Mathematics in a Real Context.* Chichester: Ellis Horwood.

Böer, H (1990). *Das Projekt Wasser.* Mühlheim: Schulpraxis.

Carr, A and Galbraith, P (1987/1991). *Practical Applications of Mathematics.* Vol 1/2. Churchill: PAM.

Clatworthy, N and Galbraith, P (1991). Mathematical modelling in senior school mathematics: implementing an innovation. In *Teaching Mathematics and its Applications,* **10**, 6-22.

Clements, R (1989). *Mathematical Modelling - A Case Study Approach.* Cambridge: Cambridge University Press.

COMAP (ed) (1985-92). *HIMAP Modules 1-19.* Arlington: COMAP.

COMAP (ed) (1981-92). *UMAP Modules - Tools for Teaching.* Arlington: COMAP

Cross, M and Moscardini, A (1985). *Learning the Art of Mathematical Modelling.* Chichester: Ellis Horwood.

Edwards, D and Hamson, M (1989). *Guide to Mathematical Modelling.* London: Macmillan.

Garfunkel, S and Steen, L (ed) (1991). *For All Practical Purposes.* New York: Freeman.

Giordano, F and Weir, M (1985). *A First Course in Mathematical Modelling.* Monterey: Brooks/Cole.

Hobbs, D (ed) (1992, 1993). *Enterprising Mathematics Course.* Oxford: Heinemann.

Huntley, I and James, G (ed) 1990. *Mathematical Modelling - A Source Book of Case Studies.* Oxford: Oxford University Press.

Jannack, W (1992). *Schalten mit Köpfchen.* Mühlheim: Schulpraxis.

Kaiser-Messmer, G (1991). Application-orientated mathematics teaching: a survey of the theoretical debate. In *Niss, M, Blum, W and Huntley , I (ed),* loc.cit, 83–92.

Kaiser-Messmer, G, Blum, W and Schober, M (1982, 1992). *Dokumentation ausgewählter Literatur zum anwendungsorientierten Mathematikunterricht.* Vol 1/2. Fachinformationszentrum, Karlsruhe.

Knauer, U (1992). *Mathematische Modellierung.* Braunschweig: Vieweg.

de Lange, J (1992). *Meaningful Math.* 3 Volumes: Matrices; Flying Through Math; Data Visualization (with Verhaage, H). Scotts Valley: Sunburst.

de Lange, J *et al* (ed) (1993). *Innovation in Maths Education by Modelling and Applications.* Chichester: Ellis Horwood.

Lovitt, C and Clarke, D (1988). *MCTP Professional Development Package:* Activity Bank. Vol 1/2. Canberra: Curriculum Development Centre.

Lowe, I (1988/1991). *Mathematics at Work: Modelling your World.* Vol 1/2. Canberra: Australian Academy of Science.

Murthy, D, Page, N and Rodin, E (1990). *Mathematical Modelling - A Tool for Problem Solving in Engineering, Physical, Biological and Social Sciences.* Oxford: Oxford University Press.

Niss, M (1987). Applications and modelling in the mathematics curriculum - state and trends. In *International Journal for Mathematical Education in Science and Technology* **18,** 487-505.

Niss, M (1993). Assessment of mathematical modelling and applications in mathematics teaching. In de Lange, J *et al (ed),* loc. cit, 41-52.

Niss, M, Blum, W and Huntley, I (ed) (1991). *Teaching of Mathematical Modelling and Applications.* Chichester: Ellis Horwood.

North Carolina School of Science and Mathematics (1988). *Introduction to College Mathematics.* 2 Volumes: Matrices; Data Analysis. Reston: National Council of Teachers of Mathematics.

Pollak, H (1979). The interaction between mathematics and other school subjects. In UNESCO (ed), *New Trends in Mathematics Teaching IV.* Paris: UNESCO, 232–248.

Spode Group (1992). *A-Level Mathematics Support Series.* Vol 1–10. Cranfield: Cranfield Press.

Starfield, A, Smith, K and Bleloch, A (1990). *How to Model it: Problem Solving for the Computer Age.* New York: McGraw Hill.

Swetz, F and Hartzler, J (ed) (1991). *Mathematical Modelling in the Secondary School Curriculum.* Reston: National Council of Teachers of Mathematics.

University of Chicago School Mathematics Project (Usiskin, Z *et al*) (1989-92). *6 Volumes: Transition Mathematics; Algebra; Geometry; Advanced Algebra; Functions, Statistics and Trigonometry; Precalculus and Discrete Mathematics.* Glenview: Scott, Foresman & Co.

2

Problems and projects for teaching modelling

Alistair Carr
Monash University, Gippsland Campus, Australia

SUMMARY

Mathematics students have traditionally been shown 'other people's models'. However, when learning to be modellers themselves, they need interesting problems likely to motivate thorough discussion and involvement. The whole modelling process requires careful thought and has many phases. Working in small groups, on extended investigations or projects, has proved to lead to effective learning of all aspects of the modelling process – and assessment should be designed to give credit for every phase.

We discuss some materials widely used in Australia, and describe ways of encouraging teachers to share with others their successes in teaching modelling.

1 THE DISTINCTIVENESS OF MODELLING

The process of inventing, developing, solving, testing and reporting on a mathematical model – that sequence of activities termed *mathematical modelling* – has been listed in many mathematics syllabi for some years now. I believe that students can be encouraged and taught to learn the skills needed in that very broad process, and that teachers are more likely to succeed in doing this if they are keenly aware of the extra dimensions that modelling activities may bring to mathematics education.

Illustrative examples of applications in mathematics have long been used in the classroom to motivate a new topic, to justify time spent in developing an algorithm, and to develop understanding via the students' own practice and analysis. These standard applications were generally presented as neat, completed models. Most aspects of the formulation and critical thinking which were at some stage crucial to the production of such a model had been hidden, or ignored, by the time it reached the textbook or classroom.

There is still a place for using standard models within a course focussing on the *process* of mathematical modelling, but teachers planning to venture into teaching modelling should be aware that they will not just be teaching applications. If teachers believe that *modelling* is just a fashionable new name for the traditional activity of sprinkling applications throughout a classroom monologue, they will do themselves and their students a disservice. Teachers embarking on teaching modelling should first experience success with modelling themselves (Galbraith, 1987).

The famous Open University seven-block diagram outlining the process of mathematical modelling is widely recommended. The phases it identifies are specifying the real problem, choosing assumptions, formulating in mathematical terms, solving the mathematical problem, interpreting the solution, validating the model, and communicating – using the model to explain, predict, decide, design, desist ... (Clatworthy and Galbraith, 1991). Mason (1989) discusses teaching methods likely to help students become modellers, such as stimulate them to ask questions, practise exploring metaphors, develop and criticise models, encourage a conjecturing atmosphere, draw attention to the movement from particular to general and vice versa. These are useful when tackling open problem-solving activities too. The skills of problem solving are used in parts of the modelling process, but modelling has an extra tangible aspect – it aims to produce a mathematical representation of a physical or social situation. Many problem-solving activities are located in the abstract realm, with no necessary link to real world events or circumstances. Many classroom teachers of modelling use class discussions, ask students to work in small groups, and choose extended investigations or projects as student tasks for class work or homework. Class discussions may continue when groups report back on their projects. Some teachers would have to modify their pedagogical practices in order to encourage students in these new learning activities, but active participation, discussion and experimentation by students are essential when they are beginning to become modellers.

It is useful to classify modelling problems by their degree of openness. Galbraith (1989) discusses three such categories. A *generalised application* introduces a modelling aspect into a standard application by altering one or more given constants, which then become variable parameters. Other ways of generalising an application may be available. From this viewpoint, a relatively routine textbook can become a goldmine of modelling opportunities. *Structured modelling* involves investigating a realistic problem context, with guiding or organising statements to chart a path to progress. *Open modelling* is done with a problem context based in reality, but with no guidance provided. There may be too much, or too little, data given. If the modelling opportunity is truly open, preconceptions about the likely mathematical approach that modellers might take should be avoided as much as possible. The process is divergent – each modeller may take a distinct approach and produce a different model, having chosen a different aspect of the real context to investigate (Galbraith, 1987). Teachers of modelling are likely to use all of these types of activity. The keys to success are developing skilled teaching methods and having a fund of stimulating problems. The experiences and successes of other teachers in finding and developing challenging modelling activities can be used.

> In the traffic lights problem it is fascinating to watch the interest and commitment grow. This first becomes evident in phase 2 of the modelling process during which participants contribute [suggested variables]. Once the group senses that it is virtually impossible to be wrong, the suggestions for potential variables come thick and fast ... experience ... indicates that this interest maintains its momentum throughout the later phases of the problem-solving process. For this reason alone the introduction of a modelling-style approach to selected problems may sustain payoff even in courses where the possibilities for its widespread adoption are small.
>
> (Galbraith, 1987)

2 PROBLEMS AND PROJECTS

Fertile problems and contexts for problem-posing are at the heart of teaching modelling. There is a wealth of such problems available (Blum, 1994). Pocket calculators have removed the need for given data to be artificial and designed to avoid long decimal numbers. Spreadsheets allow ready exploration of iterative models, or of the effects of varying a parameter's value. There is now no need for this complaint (see Barzun, 1945).

> The pupil knows the problems are not real and, since real problems are not suitably simple, the only interest that can be aroused is in the different kinds of difficulty raised by each type.

If pupils have been actively involved in the modelling process, then the simplicities will be theirs, and agreed upon, not a resented aspect of an artificial textbook question. In addition, simplifications are always potentially open to revision, in refining or revising the modelling approach taken in a particular instance.

Pioneering work in publishing modelling problems in Britain was done by Ormell (Mathematics Applicable Project, 1972-77) and the Spode Group (1981-1991). A guided modelling activity allows a particular mathematical approach to be strongly hinted at, while leaving students freedom for creative work. This can satisfy a teacher's desire to illustrate a topic or technique with an interesting application or model. Problems may be linked directly to mathematical topics, as in some of the work of the Spode Group (1986c).

A strong case for a modelling approach can be made on the grounds that, without such experience in systematic thought about real problems, students may never develop confidence in their ability to apply *any* of the mathematics they have learned. Critics of teaching modelling point out that much experience points to a one or two year gap – many students make no use of their most recently learnt mathematics when attempting open-ended modelling tasks. Other critics suspect that a course built solely as a patchwork of modelling challenges might never introduce some standard concepts and algorithms. The latter point is valid – if all the modelling activities are truly open, a teacher is duty bound to introduce new topics only when the modellers request them. At present, however, even the courses with the heaviest modelling emphasis do not resemble that patchwork scheme.

Extended project work may be used for assessment in senior secondary mathematics; an Australian experiment was described by Money and Stephens (1991). Much Australian work in teaching mathematical modelling drew inspiration from the modelling taught at the Open University in Britain. Other noteworthy features of the Open University applied mathematics programme are the use of practical or laboratory exercises, and provision of modelling data via videotape. Concrete activities are now more widely and successfully used in secondary mathematics (Lovitt and Clarke, 1988), following the pioneering work done by primary mathematics teachers. In general, there is great overlap in the types of challenges faced, and methods most likely to succeed, for upper-secondary and tertiary teachers and students of modelling.

Published sources: problems and projects

Most textbooks for tertiary mathematics present completed models and standard applications. Although this is common at the secondary level, there are also many books with a modelling flavour. An international critical bibliography by Kaiser-Messmer *et al* (1982, 1992) listed many of these, and particularly highlighted ongoing projects producing modelling materials. In Germany the group Mathematik-Unterrichts-Einheiten-Datei (MUED) has many teacher members. The Spode Group (1981-1991) in Britain has published innovative, high-quality, materials that have been influential in several countries. Some related work continues at the Centre for Innovation in Mathematics Teaching, at the University of Exeter in England. The Shell Centre at Nottingham and the Mechanics in Action Group are two other outstanding British producers. The Consortium for Mathematics and Its Applications (COMAP) in Massachusetts also has a reputation for excellence – its *HIMAP Modules*, *Consortium Newsletter* and *UMAP Journal* are used by many secondary and tertiary teachers. Malkevitch *et al* (1992) suggest many ways that modulo arithmetic and other coding principles are used in daily life, such as in bar codes, product codes and data transmission. All are within the scope of high-school mathematics.

Two highlights from Australia are Lovitt and Clarke (1988) and Lowe (1988, 1991). The former is a full professional-development package for teachers, with a strong problem-solving and modelling emphasis; the latter has a wealth of application contexts and is intended for students in the lower-ability range. These books help to dispel the notion that mathematical modelling is within the grasp of only the brightest students.

Two books by Dear (1988, 1989) from New Zealand have a wide range of open and guided modelling problems of which *Spraying the Paddock* is a good example.

> A farmer intends to spray his paddocks with herbicide. He uses a boom attached to his tractor which sprays a swathe 10 metres wide ... Assuming the farmer's paddocks are all rectangular and of varying sizes, investigate the ways of spraying may be achieved with a view to finding the most efficient.

The rate of spraying should be 400 litres per hectare, and is achieved if the tractor is driven in a straight line at 8 km per hour. Geometry and kinematics are both needed for the investigation. Dear provides teaching notes and student comments on many of the problems.

Often, teachers are careful not to say too much about a problem. One example is the problem entitled *The Far Horizon* (Carr and Galbraith, 1991):

> The distance that one can see from a point above the Earth's surface is determined by the horizon, where the curvature of the Earth limits vision. Find a formula to calculate how far a RAAF pilot can see in terms of his height above the Earth's surface.

One of the main modelling tasks here is to devise a suitable diagram, and we advise that it is "felt preferable to allow students initially to wrestle with the challenge". If a diagram were provided, the problem would reduce to another application of Pythagoras, with supplementary algebra.

In *Love Forty* (Carr and Galbraith, 1991) we set the scene as follows.

> A tennis commentator on television recently said that if a champion male player is at 0/40 when serving, he has a 50% chance of winning that game. We will try to estimate how good the player has to be for that to be true.

There follows a sequence of five questions, guiding students through probability arguments with many assumptions and calculations – but these questions could be omitted, making the bald sentences above the starting point of a more open investigation. In addition the context is fertile – many supplementary questions may occur. What of female champions? What about non-professional players? Can we collect data to test the statement? How often will the score reach 0-40 if the men are so good? Can we simulate the game? A quintic equation results from the probability argument – what methods can be used to solve such equations?

Teachers often adapt problems, for example by truncation, extension, or linking to local or current topics. Sometimes these suggestions are made by the authors – for example, shortest-path problems provide ideal ways "to use your local environment to set similar problems" (Spode Group, 1986a). Likewise network-inspection problems, generalised from Euler's celebrated Königsberg Bridge paper, can lead to "project work ... on local milk delivery, mail delivery, or newspaper rounds. You can check whether or not optimal routes are actually being used, and if not, find out why not." (Spode Group/Galbraith and Carr, 1988). Recently an American high-school teacher asked the class to investigate the new route being taken by a local recycling company, which had added two new collection points to its schedule. The students found that the route was not optimal. The company was told, it improved its route, and wrote to thank the class! These students will need no further convincing of the power and efficacy of mathematics. Project work that is locally based, and perhaps related to student concerns, hobbies, or recreations has that much more potential for fostering close involvement and enthusiasm; small-group work can help to spread the enthusiasm across the whole class, whereas large groups can still leave many participants feeling left out.

Where do good problems come from? How may we conjure up a situation whose many aspects cry out to be modelled mathematically? How do we draw up a list of case studies that students will be pleased to choose from for an extended project? Many teachers of modelling advocate telling a story, or appealing to students' altruism and idealism. Elements of mystery, detective work, helping out, or being in charge – perhaps deciding on recommendations – have their attractions. Paula Herlihy (1991) wrote three worksheets along such lines (see Carr and Galbraith, 1991)

- describing the flight characteristics of a hang glider – she was helping her brother to advertise one.
- helping a couple to design oval templates for cardboard mounts used in picture framing.
- altering a room at a Community House to use as a crèche – checking whether Department of Health requirements would be met.

Herlihy reported on the last of these.

> Morrison House is a real place, and a maths class was asked to help with the questions. The teacher's initial reaction (that it was too mundane a task for the students) was proved incorrect, as it generated much natural discussion about geometry, mensuration and trigonometry. The students took the measurements. The House made temporary alterations to meet the requirements, and then raised funds for a new building which was completed three years later.

Activities from students' daily lives, the work of their families, and current events – Olympic Games, financial news, environmental debates – may arouse interest. Teachers at in-service sessions on mathematical modelling are often asked to contribute problems to a pool for sharing. Many do not see themselves as authors, and are reluctant to expose their favourite lessons to a wider group. Yet much wisdom surely resides in the classroom experience of skilful teachers, who are able to elicit and encourage remarkable deeds and insights in their best moments and best years. A specific goal of the Mathematics Curriculum Teaching Program, coordinated by Lovitt and Clarke was to document such 'magic moments' of pedagogical excellence, to tease out their special characteristics, and to document their salient features in sufficient detail so that other teachers might emulate successful examples (Lovitt and Clarke, 1988/1992).

It should not be surprising that teachers often make valuable contributions to collections of modelling problems. Of the eleven co-authors in Carr and Galbraith (1991), for instance, three are students, three are high-school teachers, one is a secondary teacher working as a mathematics consultant, and four are tertiary teachers. This represents a fair balance of backgrounds. The Spode Group consists of eight secondary teachers, two consultants, and five academics. It meets several times a year, and members work on drafts between meetings. Another group of teachers met several times to write the estimation activities in Henry (1988). The UMAP modules and HIMAP booklets from COMAP are produced by individual authors or very small teams, with high standards set in the commissioning, editorial and publication processes. There is also the traditional dissemination via professional journals and magazines. So a variety of exemplary writing and working styles exist for publishing and disseminating modelling material. The National Educational Resources Information Service (NERIS) in England has broken new ground in producing its database on CD-ROM, and including many worksheets with possible solutions, teaching notes and suggested extension activities.

Most writers find that, once they have become attuned to the daily opportunities for devising problems in practical mathematics, there are a myriad of sources – newspapers, current events, work and recreation, government statistics, quality of life issues, science, and so on.

Assessment

Much experience in assessing modelling work has been accumulated, and published in works such as Berry and O'Shea (1982), Berry *et al* (1987), Blum *et al* (1989a). Clearly, all phases of the modelling process must be allocated credit in assessment, in order to encourage students to be thorough and self-aware in their work. It is possible to set tasks that involve only the early phases – problem definition, choice of variables, ... These can be relatively brief processes, and may be suitable as part of a traditional three-hour examination. Clatworthy and Galbraith (1991) found that the whole first year of a two-year modelling course for senior secondary students was devoted to 'preliminary training' in areas such as group work, report writing, brainstorming for suitable factors and variables, and problem formulation. Assessment criteria should be published and discussed with students, but ultimately assessment of project work involves much qualitative judgement, and complete objectivity may not be attainable. Fairness is important, but multiple-choice tests and traditional exams have many unfair aspects, and most modelling teachers believe they are quite unsuitable for testing modelling skills.

3 TERTIARY MATHEMATICAL MODELLING

The teaching of modelling – as a process – is still not widespread. Many applied mathematics lecturers would assume that students will imbibe good methods by sitting at the feet of the greatest modellers, from Newton and Euler to Gauss, Maxwell, Heaviside, Stokes and Helmholtz. Examples of successful modelling courses are described in Berry *et al* (1987) and Blum *et al* (1989b), for example.

At my institution, we pay attention to the skills of modelling in three courses. *Mathematical Modelling A* was developed from a first-year applied mathematics course, linked to calculus and focussing on physical systems. It now has a strong modelling component comprising about 50% of class time and 60% of the assessment credit. A segment of *Scientific Thought and Methods*, a subject for all second-year Applied Science students, is devoted to introducing modelling and is linked to other thinking skills discussed in that course. Mathematics majors may enrol in *Mathematical Modelling B* in the third year. This develops their modelling further, one assignment asking students to take the list of thirty-six characteristics of mathematical modelling in Kapur (1988) and to illustrate twenty-five of them by mentioning at least a dozen models, sub-models or modelling examples. This task focusses attention on the generic aspects of modelling earlier in the course, and encourages students to survey the mathematics, and other areas, they have learned since their introduction to modelling. The technical component of the course is based on Beltrami (1987), which covers a rich array of models in dynamic systems theory, with good starting points for project work. The basic textbooks for these courses are Giordano and Weir (1985) and Edwards and Hamson (1989). Both texts are suffused with an awareness of the challenges and delights of modelling, and are written at a suitable level of mathematical sophistication. Our students, some of whom are distant students, are provided with printed study guides.

I have found the periodicals *Teaching Mathematics and its Applications* (The Institute of Mathematics and its Applications), *International Journal of Mathematical Education in Science and Technology*, and the *UMAP Journal* and *UMAP Modules* (both COMAP) very useful.

4 CHALLENGES

There are still opportunities for a modelling flavour to 'seep' into other tertiary mathematics units, so that it is not seen as a 'separate country' – I believe that modelling can motivate the learning of all mathematics. Adaptation of published problems is the general rule, not the exception, and should be encouraged if teachers are becoming more involved in re-assessing their classroom practices.

The question of the ultimate effectiveness of using modelling in mathematics education is still open, as Kaiser-Messmer (1993) points out. Teachers and many students are enthusiastic, but there is still a dearth of empirical studies of its ultimate effects on learning. Does success depend heavily on a teacher's personality? If so, can adoption of a modelling ethos actually change the *persona* that a teacher presents in class, and/or the classroom atmosphere, to such an extent that more creative and enduring learning will occur?

REFERENCES

Barzun, J (1945). *Teacher in America.* Boston: Little Brown.

Beltrami, E (1987). *Mathematics for Dynamic Modeling.* Boston: Academic Press.

Berry, J S *et al* (ed) (1987). *Mathematical Modelling Courses.* Chichester: Ellis Horwood.

Berry, J S and O'Shea, T (1982). Assessing mathematical modelling. In *International Journal of Mathematical Education in Science and Technology,* **13**, 715-724.

Blane, D and Evans, M (ed) (1989). *Mathematical Modelling for the Senior Years.* Parkville: Mathematical Association of Victoria.

Blum, W, Niss, M and Huntley, I (ed) (1989a). *Modelling, Applications and Applied Problem Solving.* Chichester: Ellis Horwood.

Blum, W *et al* (1989b). *Applications and Modelling in Learning and Teaching Mathematics.* Chichester: Ellis Horwood.

Blum, W (1994). Mathematical modelling in mathematics education and instruction. In Breiteig, T, Huntley, I and Kaiser-Messmer, G (ed), *Teaching and Learning Mathematics in Context.* Chichester: Ellis Horwood, 1-12.

Carr, A and Galbraith, P (ed) (1991). *Practical Applications of Mathematics. Vol 2.* Churchill: Project PAM.

Clatworthy, N and Galbraith, P (1991). Mathematical modelling in senior school mathematics: Implementing an Innovation. In *Teaching Mathematics and its Applications,* **10**, 6-28.

Dear, R (1988). *Problem Solving of the Third Kind.* Invercargill: Southland Mathematical Association.

Dear, R (1989). *Modelling Activities for the Classroom.* Wellington: New Zealand Mathematical Society/New Zealand Association of Mathematics Teachers.

Edwards, D and Hamson, M (1989). *Guide to Mathematical Modelling.* Basingstoke: MacMillan.

Galbraith, P (1987). Modelling – teaching modelling. In *Australian Mathematics Teacher,* **43**, 6-9.

Galbraith, P (1989). From applications to modelling. In Blane, D and Evans, *op.cit.*, 78-86.

Giordano, F and Weir, M (1985). *A First Course in Mathematical Modeling.* Monterey: Brooks/Cole.

Henry, B (ed) (1988). *Guess, Estimate, Measure.* Parkville: Mathematical Association of Victoria.

Herlihy, P (1991). In Carr, A and Galbraith, P (ed), *op.cit.*

Kaiser-Messmer, G, Blum, W and Schober, M (1982, 1992). *Dokumentation ausgewählter Literatur zum anwendungsorientierten Mathematikunterricht.* Vol 1, 2.

Kaiser-Messmer, G (1993). Reflections on future development in the light of empirical research. In Breiteig, T, Huntley, I and Kaiser-Messmer, G (ed), *Teaching and Learning Mathematics in Context.* Chichester: Ellis Horwood, 213-218.

Kapur, J N (1988). *Mathematical Modelling.* New Delhi: Wiley Eastern.

Lovitt, C and Clarke, D (1988/1992). *Mathematics Curriculum Teaching Program – Professional Development Guides.* Carlton: Curriculum Corporation.

Lovitt, C (1992). The professional development of teachers through recognising, documenting and sharing the wisdom of practice. Québec: Lecture at ICME-7, Québec.

Lowe, I (1988, 1991). *Mathematics at Work: Modelling Your World.* Vol 1, 2. Canberra: Australian Academic of Science.

Malkevitch, J, Froelich, G and Froelich, D (1992). *Codes Galore. HIMAP Module 18.* Arlington: COMAP.

Mason, J (1989). Modelling: What do we really want students to learn? In Blane, D and Evans, M, *op.cit.*, 7-21.

Spode Group (1981, 1982). *Solving Real Problems with Mathematics.* Vol. 1, 2. Cranfield: Cranfield Press.

Spode Group (1986a). *The Decision Maths Pack.* London: Edward Arnold.

Spode Group (1986b). *Decision Mathematics.* Chichester: Ellis Horwood.

Spode Group (1986c). *Motivating A-Level Mathematics.* Oxford: Oxford University Press.

Spode Group (1989). *The Active Maths Pack.* London: Hodder & Stoughton.

Spode Group (1991). *The Applied Maths Pack.* London: Hodder & Stoughton.

Spode Group/Galbraith, P and Carr, A (1988). *Decision Maths Pack – Australian Edition.* Caulfield East: Edward Arnold (Australia).

3

Conceptions of mathematical modelling and their implications for the future

Zalman Usiskin
University of Chicago, USA

SUMMARY

The future of mathematical modelling in elementary and secondary classrooms depends to a great extent on the perceived distance between modelling and other mathematical activities which may take place. This paper discusses various conceptions of the relationships between pure and applied mathematics held in the mathematical community, and the consequences of each.

INTRODUCTION

At the 6th International Congress on Mathematical Education, ICME-6 in Budapest, Werner Blum and Mogens Niss gave a summary of the then-present state of mathematical modelling (Blum and Niss, 1991). Judging from that paper, I believe that it can be said that there has been a great deal of work in many countries on mathematical modelling, and probably more work since the last ICME than in any other four-year period ever before – but what will happen as a result of this work? This paper considers various conceptions of mathematical modelling and suggests some implications of these conceptions for the teaching of mathematics as a whole (see also Kaiser-Messmer, 1991).

CONCEPTION 1:
Applied mathematics is paramathematics

In the 1960s, at the time of New Math, when in the United States, France and many other countries the curriculum was devoid of applications, those who called for an increased emphasis on applications were viewed as outsiders who did not really understand what mathematics was about (Kline, 1958; Meder, 1958). Applied mathematics was viewed as inferior mathematics, and applied mathematicians as inferior mathematicians. When a few applied mathematicians with substantial credentials as mathematicians called for applications, a common view was that they were biased by their professional concerns. We might say that applications were viewed as *paramathematics*, material that uses good mathematics but involves inferior problems and techniques, in the same sense that a paramedic is obligated to use the best principles of medicine but is not considered to be the equivalent of a doctor.

There are still some who feel this way towards applications of mathematics (see, for example, Halmos, 1985). Many teachers do not want to teach applications or modelling and, in fact, some went into mathematics precisely because they wanted to avoid the real world. I know this because I myself was one of those teachers.

Most members of the mathematics education community nowadays believe that applied mathematics is at least on an equal footing with pure mathematics, but we must be aware that many others do not share our beliefs. We are the flock and we are talking to each other. There are still those who believe that the more abstract an idea is the more widely applicable it will be.

CONCEPTION 2:
Applied mathematics is when you use something from one branch of mathematics to study something from another branch

There are still those who believe that applications internal to mathematics mean the same thing as applications that originate in the real world. In this way of thinking, an application of geometry is when the Pythagorean Theorem is used to develop the formula for the distance between two points (x_1,y_1) and (x_2,y_2) in the coordinate plane. Modelling is when one builds polyhedra or uses the Klein or Poincaré models of non-Euclidean geometry. We need to educate our colleagues, and to do so we will have to use their language, not ours.

CONCEPTION 3:
Applied mathematics is a service to those who will not study mathematics

In a recent seminar on the question of geometry in the scientific education of students in grades 11-18, one of the four reasons given for teaching geometry was its rôle as a service for students who will study geography, crystallography, robotics, and other disciplines. This is of course true, but I think most of us tend to think of these other subject areas as having as much potential to serve 'us' in the teaching of mathematics that is, to provide contexts for the teaching of problems, methods, and concepts as to serve 'them' as a place where mathematics is applied.

CONCEPTION 4:
Applied mathematics is the way to approach mathematics for those who cannot do pure mathematics

This is somewhat a corollary of the first and third conceptions. The idea is that we should teach applied mathematics to some students because we do not have much choice with these students. They really *need* mathematics, but they are not successful at the best mathematics, so let us teach them applications.

CONCEPTION 5:
Applied mathematics is only for the best students

In many places students do not see applications of mathematics beyond arithmetic until they are at the tertiary level, where there are fewer and brighter students. In some places in the United States, the only younger students who do more than a few applications are those who have finished the teacher's syllabus – the teachers feel that a student must 'know' the mathematics before applying it. Sometimes this is put in the context of first learning the basics – which in this conception do not include applications – before learning anything else. Applications and modelling are for the day before Christmas or the warm days in spring towards the end of the year.

The combination of conceptions 4 and 5 means that, in many places, typical students do not have many experiences with applications and modelling, while those who are either quite poor students or quite good ones do, on occasion, have such experiences. Another factor causing this result is that the curriculum for typical students (the 'average', the ones in the middle) is more difficult to change than a curriculum for special populations.

CONCEPTION 6: Applied mathematics is fundamentally different from pure mathematics

I have begun to hear of this conception more and more in recent years. This conception begins with many beliefs, among them the following. Applications are inherently messy (see Halmos, 1985). Applied mathematics is inductive. In applications, problems usually have many answers. Algebraic and arithmetic skills are not as important in applications, for they can be done by calculators and, anyway, exact answers are not what we want – estimates are good enough. Applications and modelling involve discovery and exploration. In applied mathematics, it's the process that is important, and particularly the attitudes of the students.

The difficulty with this view is that it is also true of pure mathematics. Much of pure mathematics is also messy, from infinite decimals and long division to long summations and integrals. We often conjecture inductively in pure mathematics. Proofs and other problems have many solutions and often many answers. In the time of new math, when applications were not in the curriculum, did we not hear of how unimportant algebraic and arithmetic skills were? Is not estimation and error analysis a part of pure mathematics? Does not pure mathematics involve discovery and exploration, and are not process and attitudes important there also?

Furthermore, are these characterisations true for all applications? When we go to a shop, we have many options regarding what to buy, but given our choice there is one answer to what we will buy. An estimate will not suffice. One of the major reasons that mathematics is so often used by people outside mathematics is that it is not messy, and it does give a single answer in so many situations. There is content in applied mathematics, too. For example exponential growth, rate of change, various counting principles, the mensuration formulas, the path of a projectile, and so on. Are not deductions from assumptions one of the major aspects of the modelling process? I believe that applied mathematics and modelling are not all process, any more than pure mathematics is all product.

CONCEPTION 7: Applied mathematics parallels pure mathematics

In my own mathematics education, like that of many of my colleagues, there were few applications and no mathematical modelling at all. My conversion to think of applications as necessary components in a mathematics curriculum took place in the early 1970s; the work of my colleague Max Bell was instrumental in this conversion. By ICME-3 in 1976, when Henry Pollak spoke of the similarity of pure and applied mathematics (Pollak, 1979), I had come to the same view. Those who believe this conception find it quite natural to move from one to the other and, on occasion, to be in a suspended state in which one does not know whether one is dealing with pure or applied mathematics (see Lax, 1990).

There is, however, a fundamental difference in structure between pure and applied mathematics. Pure mathematics *has* structure. In fact it has two kinds of structures. One kind of structure is the well-known mathematical system, the structure of postulates, definitions, and theorems related by logic and deduction. This structure has been developed over thousands of years and is utilised not only in our organisation of mathematics but also in our teaching, particularly in geometry and upper-level courses.

In this structure, the basics include such properties as the commutative and associative properties of addition and multiplication, and the Triangle Inequality. The typical activities we have students do are expressed by the verbs *justify* or *prove*. The highest level of thinking for the student is the proving of a theorem the student has not seen before. The highest level for the mathematician is the discovery of the proof of a theorem that no one has seen before.

A second structure of mathematics is algorithmic in nature. It is the structure that we often apply to the learning of manipulative skills. For instance, with paper and pencil one cannot do long division (either with numbers or polynomials) unless one can multiply and subtract. This algorithmic structure has long been used to organise school arithmetic and algebra. The basics of this structure are the memorised facts such as $7 \times 8 = 56$, or the construction of the perpendicular bisector of a segment or the solving of the equation $ax + b = c$. The typical student activities in this structure are expressed by the verbs *find* or *simplify* or *invent*. The highest level of thinking for the student or mathematician is to invent a new way of finding an answer to a problem – to invent a new algorithm.

Jan de Lange recently identified several problems facing problem-oriented curricula such as those commonly put forth as appropriate for applied mathematics. Among these problems were a loss of basic skills and a loss of structure. I think these losses derive from the fact that, for problem-solving curricula, we have not yet developed a structure comparable to the mathematical systems and algorithmic structures of pure mathematics, and I think we should treat these comments as not only defining a problem but also providing an opportunity. The problem is that we should make certain that, in our enthusiasm for rich applications, we do not ignore the structures of pure mathematics that have been developed over the centuries. The opportunity is that we should work to develop similar structures for applied mathematics. We have tried to do so in the materials of the University of Chicago School Mathematics Project for secondary school students (Usiskin, 1989).

Of course, there is not just one mathematical system, not just one way to order the theorems of mathematics. Nor is there just one way of ordering the algorithms of mathematics. Indeed, technology has caused all of us to reexamine the algorithms we use in mathematics, and suggests major reorderings of topics in school mathematics. Likewise, there are many possible structures for applied mathematics.

In the structure for applied mathematics we already know the highest form of thinking for students. It is for them to use a mathematical model to attack successfully a real problem, or to understand a real situation they have not previously studied. This is akin to proving a theorem they have never seen before. At the highest level, the applied mathematician develops models no one has ever seen before. Typical student activities in dealing with applied mathematics are *model*, *optimise*, and *predict*.

There are many similarities between proving and modelling (Usiskin, 1991). To work with proofs or models in areas not understood by students is likely to be a waste of time. Proving or modelling in one domain does not necessarily extend to the proofs or models in another domain. Both proving and modelling require the language of if-then statements and of deduction. Communication of proofs and models both require that students write using both mathematical and non-mathematical terminology; the writing task itself poses difficulties for many students. Competence in proofs and in forming mathematical models comes slowly.

So, at the highest end, there is a wonderful commonality between pure and applied mathematics. What about at the lowest end? What are the basics of applied mathematics?

To many, these basics are processes such as estimating, sorting out relevant from irrelevant information, and organising information. However these processes are unsatisfactory as basics for applied mathematics, for they apply as much to non-applied problems as they do to applied mathematics. I believe the basics are to be found in the connections between mathematics and the real world, in identifying the fundamental real-world ideas that are manifested in mathematical ideas such as number, the various operations on number, the various functions and operations on them, the manifestations of point, line and various curves. They are the simplest forms of the content of applied mathematics.

CONCEPTION 8:
Applied mathematics is mathematics

More and more, in trying to sort out issues of curriculum and instruction with respect to modelling and applications, I come to the conclusion that, for the development of applied mathematics in the curriculum to continue to improve, there must be some realisation that much of what we talk about as process in the teaching of applications and modelling – the immersion of students in activities, the provision of choices of what to do, asking them to explore – is simply good teaching. There must also be some realisation that the connections that are made between mathematics and the real world are as much a part of mathematics as the connections that are made between algebra and geometry. The similarities between pure and applied mathematics are not coincidence; they are there because these are both an essential part of the magnificent structure we call mathematics.

CONCLUSION

The future of mathematical modelling in elementary and secondary mathematics classrooms depends, to a great extent, on the perceived distance between modelling and other mathematical activities which may take place. The various conceptions of the relationships between pure and applied mathematics discussed here range from those in which mathematical modelling is viewed as a distant departure from the teaching of mathematics to those in which mathematical modelling is viewed as an integral part of the teaching of mathematics, and thus of zero distance from it. It is the opinion of the author that it is most advantageous to take the latter of these conceptions, but to do so without taking into account that many people believe the other conceptions is likely to be counterproductive.

REFERENCES

Blum, W and Niss, M (1991). Applied mathematical problem solving, modelling, applications, and links to other subjects – state, trends and issues in mathematics instruction. In *Educational Studies in Mathematics*, **22**, 37-68.

Halmos, P (1985). In Albers, D J and Alexanderson, G L (ed), *Mathematical People*. Boston: Birkhäuser, 119-132.

Kaiser-Messmer, G (1991). Application-orientated mathematics teaching: a survey of the theoretical debate. In Niss, M, Blum, W and Huntley, I (ed), *Teaching of Mathematical Modelling and Applications*. Chichester: Ellis Horwood, 176-181.

Kline, M (1958). The ancients versus the moderns, a new battle of the books. In *Mathematics Teacher*, **51**, 418-427.

Lax, P D (1990). In Albers, D J, Alexanderson, G L and Reid, C, *More Mathematical People*. Boston: Harcourt, Brace, Jovanovich, 139-160.

Meder, A E (1958). The ancients versus the moderns, a reply. In *Mathematics Teacher*, **51**, 428-433.

Pollak, H (1979). The interaction between mathematics and other school subjects. In *New Trends in Mathematics Teaching IV*. Paris: Unesco, 232-248.

Usiskin, Z (1989). The sequencing of applications and modelling in the University of Chicago School Mathematics Project (UCSMP) 7-12 Curriculum. In Blum, W *et al* (ed), *Applications and Modelling in Learning and Teaching Mathematics*. Chichester: Ellis Horwood, 176-181.

Usiskin, Z (1991). Building mathematics curricula with applications and modelling. In Niss, M, Blum, W and Huntley, I (ed), *Teaching of Mathematical Modelling and Applications*. Chichester: Ellis Horwood, 30-45.

Section 2

Examples and issues

4

Exploring numbers through subtraction in primary mathematics

Dagmar Neuman
Department of Education, University of Gothenburg, Sweden

SUMMARY

In a two-year teaching experiment with first- and second-graders, it was illustrated how children, through communication and concrete actions, can create their own mathematics. The teaching experiment began as a game. The children were supposed to live in *Long-ago Land*, where there were no counting words, no figures, no mathematics at all. It aimed at creating experiences through which children's own informal ways of understanding mathematics could be developed in what was assumed to be a 'natural' way, according to interpretations made in interviews with school starters. This way was in some respects suggestive of the way in which our decimal system was created.

1 BACKGROUND AND AIM

Unlike the case of addition, at least two primitive behavioural ways of thinking about subtraction have been described in research on early addition and subtraction: *Take away* (John has 9 kronor. He drops 7. How many kronor are left?) and *Building up* (John has 2 kronor. How many more does he need in order to buy something that costs 9 kronor?). The 'take away' model is intuitively interpreted as subtractive and the 'building up' model as additive (Carpenter and Moser, 1982). The problem children have in dealing with subtraction is that these primitive models interact with number size in a way that sometimes severely complicates problem solving.

Pupils with maths difficulties count and keep track

Interview studies of Swedish pupils with maths difficulties (Neuman, 1987, 1989a) revealed that, even after 10 years at school, many were not able to solve the above problems through mental calculation. They listed 7 words, backwards in the subtractive and forwards in the additive problem, while they kept track of the words with their fingers. Since the semantics of the problems always ruled their choice of strategy, they only found subtraction easy if the 'taken away' or 'missing' part was within the subitising range.

School beginners expect maths problems to be easy

In an interview study (Neuman, 1987, 1989a, 1989b, 1992), where 105 school starters, aged 7 were given verbal addition and subtraction problems within the number range 1–10, some children revealed a way of dealing with this kind of difficulty. They used their fingers in the same way as we once must have used the ancient Roman numerals, which from the beginning were pictures of our hands, with 9 written as VIIII and 4 written as IIII. For the problems above they often immediately, without counting, put up a 'finger-number' starting at the little finger of the first hand and ending at the ring finger of the second, or they just looked at their ten fingers, knowing (and, when asked, telling about) where the 'ninth finger' (or as they said 'nine') was situated. They seemed to avoid splitting up the fingers of their first hand, and could thus see both the kinds of subtraction problem above in the same way. In problems where both parts were less than five, the children would move the thumb, or sometimes a couple of fingers, from one hand to the other.

Through the 'undivided five', constituted through the first hand which was not split up, the parts within the whole of the ten basic numbers of our decimal system became easily perceptible gestalts. The larger part within these gestalts had to be the first one, related to the 'undivided five'. Thus, it was ordinally experienced, since – in the example used here – it ended at the 'seventh-finger'. However, simultaneously it was cardinally experienced, since it was perceived as '5 + 2'. In the same way the whole - 'nine' - was experienced as an inseparable ordinal-cardinal number. The smaller part, which then became the last one, can never be ordinally experienced. Yet, it is small enough to be cardinally experienced without any grouping. When numbers were perceived in this way, the semantics of the problem no longer impeded functional problem solving (see Fig. 1, where the 'finger numbers' are replaced by Roman numerals).

Figure 1

It should be stressed that, if fingers were used, they were always used as 'finger numbers' in 'take away' subtraction, often used in 'building up' subtraction, and sometimes used in addition. However, in addition the fingers were also used in many other ways. Thus, the easily perceptible part-part-whole gestalts of numbers mostly seemed to be created through subtraction.

Some children had to count on their fingers sometimes to know where a 'finger number' ended; others tried to count without using their fingers. Yet the complicated counting and keeping-track strategies used by the older pupils with maths difficulties were rarely used.

On the contrary, the school starters seemed to expect that maths problems should be easily and instantly solved. Pupils who were a bit more advanced than those forming concrete 'finger numbers' could explain their immediate and correct answers, saying that they 'thought with their hands'. The most advanced pupils could even explain the instantly-given answer 'Seven' to the 'building up' subtraction above by saying 'Took away two', and explain the answer 'Two' to the 'take away' subtraction by saying 'Because two plus seven makes nine'. Thus, they subtracted in the same ways as their class mates who used concrete 'finger numbers', not bothering about the semantics of the problem. Their answers seemed to have been given through so called 'known number facts', in spite of the fact that these children had neither been drilled on 'basic facts' nor received formal teaching of any kind.

Other school beginners had, on the other hand, not fully understood the meaning of the counting words, and expressed very early conceptions of number. However, the expectation that 'answers to maths problems should be immediately experienced' was expressed even by them. They could, for instance, answer 'Nine' to the 'building up' subtraction above, behaving as if they were measuring the number of pencils with the number word sequence, telling the interviewer that the number of the unknown part ended at the word 'nine' in the sequence. They were not yet aware of the contradiction between 'nine' simultaneously used for denoting this part and the whole. Other children estimated. Their answer to the problem was of the kind 'Six ... or seven ... might be eight ...you can't know ...'. They seemed to think of the number words as synonymous with 'a little', 'something in between' and 'much', well understanding that if the known part of pencils was large, the unknown had to be small.

The children who estimated and also those who denoted the unknown part using its 'limit name', seemed to experience the nine pencils as a continuous quantity, not divided up into units. This illustrates an early experience of number. Piaget (1969) has, for instance, emphasised that children who do not understand that a number is a number of units do not have a fully developed conception of number, and that all his data point to the sequence of whole numbers as being inseparably ordinal and cardinal. These children did not express experiences of ordinal-cardinal numbers, but of 'extents', or of 'extents' ending at a certain point of the number word sequence.

A teaching experiment, carried out with two of the interviewed classes, took the information gathered in this study as its starting point.

2 THE PHENOMENOGRAPHIC RESEARCH APPROACH

A phenomenographic research approach (Marton, 1981) was used in the teaching experiment as well as in the two-interview studies. Phenomenography regards knowledge to be man-world relations created through our experiences. Learning is seen as change of conception, caused by challenging problems or conflicting experiences. The aim of phenomenography is to reveal conceptions of different phenomena, created through experiences. This is mostly done through individual clinical deep-interviews, which are tape recorded and transcribed verbatim. Teaching experiments are conducted, to explore whether teaching situations can be arranged through which pupils can get the kind of experiences that has been interpreted in the preceding interview study as the roots of informally created knowledge, and if these experiences then give rise to a similar kind of knowledge.

3 THE TEACHING EXPERIMENT

The aim of the teaching experiment (Neuman, 1987, 1989) was to create a teaching situation where the early ways of understanding number, subtraction and addition could be further developed in the way in which the interviews with school starters indicated it might be developed. Of course, the need for further development for the more advanced pupils in the classes was not ignored.

Design of the study

Two classes took part in the study, which began when school started and continued to the end of grade 2. Both these classes had been interviewed at the start of school, as had a further 23 pupils in five other classes, comparable with the research classes with respect to social environment. These 23 pupils became a control group. About 50% of the 23 children were thought to be pupils with potential maths difficulties, according to a simple test carried out before the interview study. At the end of the second year, the pupils who had no correct answers on the subtraction problems given in the interviews at the start of school were interviewed again, both in the research and the control classes. These pupils were the ones with the earliest conceptions of number.

The Long-ago Land

About 15% of the children in the two classes, when they began school, seemed to experience numbers as some kind of 'extents' or continuous quantities not divided up into units. The number words seemed to be confusing for these children.

Therefore the teaching began with a game, where the children were supposed to live in the Long-ago Land. There were no counting words, no figures, no measures, no money, and no schools with a subject called maths on the timetable. There was no maths at all, but sometimes problems 'popped up', demanding different forms of mathematics. Then the people in the Long-ago Land had to invent the mathematics they needed, as the occasion demanded.

To experience the need to split a continuity into units

The very first lesson began with two servants coming to the King and complaining about the unfair salaries his chamberlain had given to them, in spite of the fact that they had done exactly the same work. Both had got 'golden sand' in a vessel as a salary. One of the vessels was a tall, narrow bottle and the other a low, wide pot. 'Who had reason to be dissatisfied?' asked the teacher. Most children thought it was the one who had got the pot, because that was 'so small'. A few, however, pointed to the bottle saying that it was very narrow, so there might be less gold in that one. 'But how can I know?' asked the King (the teacher). 'Can anyone help me find out if my chamberlain has been unfair or not?'

Some children proposed that one could measure. The teacher reminded them that there were no measures in the Long-ago Land. Yet the children thought they could use some objects they found on a table full of things: glasses of the same and of different sizes, spoons, bits of yarn, square tiles, sticks of different sizes, and so on. The children chose, with great accuracy, just the right glasses as measures – neither too large nor too small. When all sand was poured out there were three full glasses beside each vessel. Then the King said:

> Here you see, there were as many units in the bottle as in the pot. Then, we can put this sign between the vessels to show that the chamberlain has been fair. We will call it an 'sign og equality', since there are always as many units on one side of such a sign as on the other. (Draws the sign of equality 'in the sand', that means on the blackboard, where he earlier had drawn the picture in Fig. 2.)

Figure 2

Neither the word 'unit' nor the word 'equals sign' was explained. A word and a written symbol related to an experience was what this lesson intended to give the children, who were to have many more experiences of units and of the equals sign during the lessons which followed. The next time, for instance, there was only one glass of each size. The children then found out that this could be used many times. This time, 'units' meant piles of sand, while the glass stood beside as a measure.

To experience the need for number symbols

Some weeks later, the servants were paid with oil (water), which they needed for their lamps. Again, there was only one glass of each size, but as each unit of 'oil' had to be poured into a bucket in the measuring procedure this time, the units 'disappeared'. Then the need for number symbols became urgent. 'How can we keep track of all the units that are being poured out?' asked the King. The children, as usual, found a solution. 'We can put up one finger for each of them', some of them proposed, while others thought they could draw strokes 'in the sand' (on the black board). Thus, precursors of 'finger numbers' and 'Roman numerals' were created.

Different kinds of units, and the plus sign

In due course, different kinds of continuous quantities were measured: cloth, yarn, sticks, and so on. The cloth (large pieces of painted paper sheets of different shapes) was supposed to be valuable silk brocade with golden threads and diamonds, left over from the dresses which the queen's two servants had made for her. Each girl had been given one piece of silk brocade in payment, but they thought they had not been fairly treated. The pupils now knew that without measuring they would not be able to know if the two pieces of silk were of similar size or not. However, three of the four groups of children behaved this time in an unexpected way. They took the pieces of paper sheet and stuffed them into two large glasses, saying:

> Look! It's the same: If you press really hard, it just fits – and there's just as much in both glasses!

For them, then, 'measuring' was to 'see how many glasses' could be filled. This way of measuring could not be seen as wrong. Yet, the aim of this lesson was to help the children understand that different kinds of measures had to be used for different kinds of quantities, and also that measuring is to separate one unit from the whole to see how many times it fits into the whole. Therefore, the next day there were two new pieces of valuable silk brocade, but this time of different sizes. Now one piece was too large to fit into the glass, and the children understood that the chamberlain had not been fair to the girls. The Queen (the teacher) suggested that the silk should be put back, and asked the children if they could think of some other way of measuring it. After one unsuccessful attempt to measure with a stick, one pupil finally found a tile which (for some reason!) fit exactly between the sides of the cloth. Yet, when measuring, she put her hand as a 'border' every time she moved the tile. The teacher asked how she would measure the cloth under her hand. The children then decided to work together: one holding the tile, another using a chalk for drawing limits where each unit ended, and a third writing one stroke for every unit marked, while the other children in the class raised a finger for every one. They easily saw that there were as many units as the hand in one piece, and as many as the hand without the thumb in the other.

The queen cut one unit more from her silk and took two 'prickly leaves' (two pieces of green tape) to fasten the missing unit to the piece where it belonged, and the pupil at the blackboard added one stroke for the new unit.

The queen then said:

> You have forgotten something. You have not drawn a cross for the leaves I used for fastening the unit.

The pupil wrote a '+' before the last stroke, and the queen continued:

> I think these leaves, and the cross you wrote, should have a name. We will call it 'plus'. From now on, if we use such a 'plus' sign, we can think of it as 'belong together' or as 'add on' to something (Fig. 3).

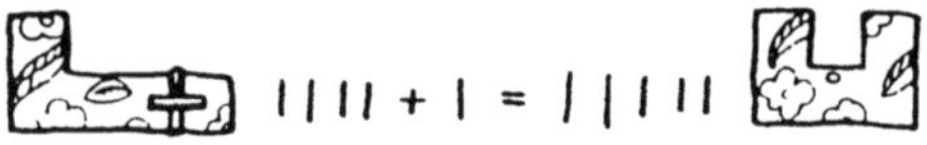

Figure 3

The first number sentence had been written. It described a 'building up' subtraction. Subtraction must be the first counting operation if children are thought to acquire understanding of number through comparison of quantities.

Discrete quantities and the need for grouping

Some weeks later discrete quantities appeared: spoons, sheep and so on. On one occasion the King decided that each family in his country would have to pay a tribute, one sheep for each family. Many sheep arrived, and the treasurer wrote stroke after stroke after stroke in the book where everything was accounted for (Fig. 4a). There were lines and lines of strokes in his book, but no one could actually work out exactly how many sheep they represented. The treasurer also got tired of writing all these strokes. What could he do?

As usual the children had suggestions: 'When we play games at home we write strokes like this' (Fig 4b) one boy exclaimed. The King said 'Yes, as one hand with the thumb folded over it'. This helped a bit, but the treasurer still got just as tired, because there were still as many strokes to draw. The King then proposed that the hand could be written in a simpler way, as V, one stroke for the fingers held together and one for the thumb (Fig 4c). When there were as many strokes as the fingers of both hands, an *X* could be written: two hands, one of them upside down. The children used their both hands with the thumbs crossed to illustrate *X*.

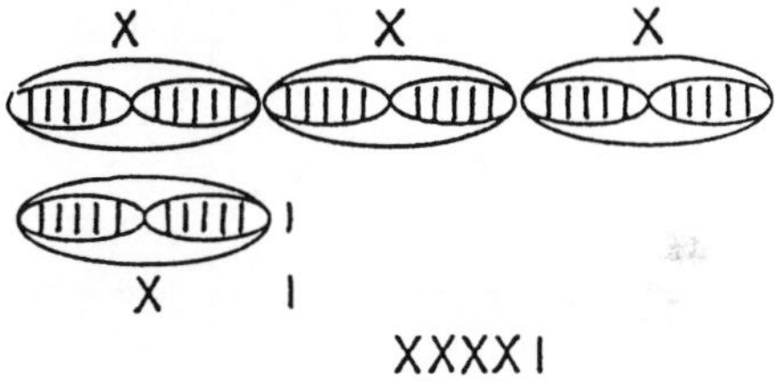

Figure 4a

Figure 4b

Figure 4c

After that each child received a piece of paper with all the chamberlain's strokes written on, and were encouraged to rewrite the strokes as hands and single fingers. Then it was simple to perceive the number of sheep that had come to the castle.

When the discrete quantities were measured the size of the units was not interesting: 'one sheep' was one unit, independent of how big or how small it was.

Moving out of the Long-ago Land

The fairy tale went on until the Roman numerals had been exchanged for our Arabic figures, and the counting words had been introduced. The idea of place-value was a thrilling experience in this context. It was easily understood when related to the Roman numerals.

The problem-solving approach was retained when the children had 'moved out' of the Long-ago Land. The new problems could be aimed at reflecting on ordinal numbers, how they differed from and were related to cardinal numbers, and which numbers were possible to divide up into parts, and so on. Transformations appeared in many contexts, for instance in problems of the kind exemplified in Fig. 5.

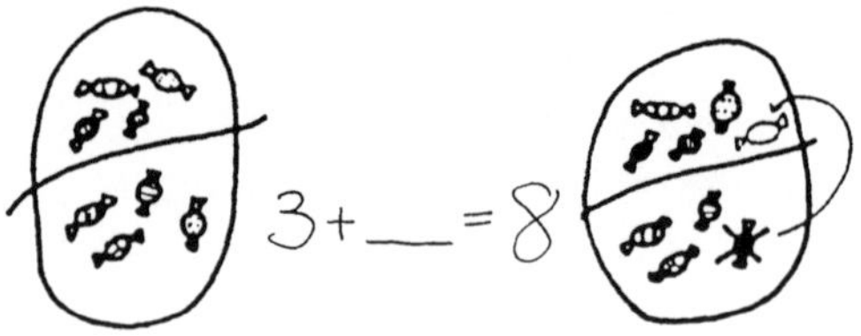

Figure 5

The idea of an 'undivided five' was also created in many contexts, for instance when 5-kronor coins and 1-krona coins were used. Yet, it was introduced through finger numbers related to Roman numerals.

The children had never been *presented* with number sentences written by the teacher during their first months at school, even if they themselves had written a lot of number sentences to describe their own activities. However, towards the end of the first term the children were presented with the number sentence

$$9 = \_ + 2$$

written by the teacher on the blackboard (Fig. 6a).

It was a 'building up' subtraction where the added number was large, and it was expected to be a real problem for quite a lot of children. The teacher said:

> Yesterday I had some sweets in my pocket, and I gave some to Christine. Here, on the blackboard, I have written down how many sweets I myself have and how many I had from the beginning, but I do not remember how many Christine got. Can anyone say how many I had from the beginning? And how many I have in my pocket now?

The children, who had written a lot of number sentences describing different actions, could easily answer those questions, and the teacher went on:

> Then, can anyone tell me how many sweets Christine has?

Some children found the solution directly. Others were sitting quietly, pondering, and the teacher said:

> I see, there are some of you who find this difficult ... Then ... why do you not use the 'mini-calculator' they used in the Long-ago Land?

The pupils looked confused. Mini-calculator in the Long-ago Land? As far as they knew, nothing like that had existed there. The teacher held out her hands saying:

> This one! Show me with your hands how many sweets I had from the beginning! Can someone also go to the blackboard and write down the Roman numerals for that number?

VIIII was written below 9 in the number sentence. After that the teacher asked the children to circle the number representing the sweets she had left in her pocket. When that was done it was easy to 'read off' the number denoting the number of sweets given to Christine (Fig. 6b).

Figure 6a

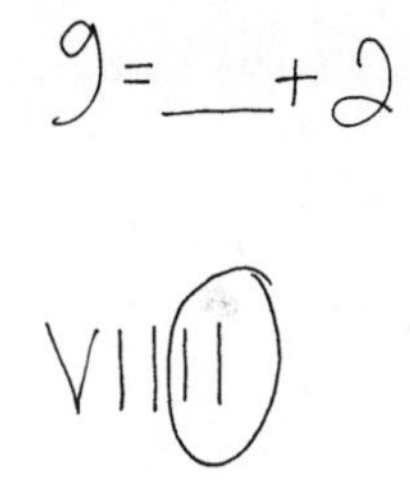

Figure 6b

The numbers had now become inseparably ordinal-cardinal, and could be experienced as 'gestalts', where the parts within the whole were immediately perceptible.

The experiment went on ...

It is not possible to describe all problems that were worked out during the two years the teaching experiment went on. 'Take away' subtraction, with the minus sign, was experienced as 'even easier than plus' as the children expressed it after having tried these kinds of problem, and neither was addition – when finally presented – experienced as different.

During the third term of the year, all the children could immediately – by 'thinking with their hands', or thinking in even more abstract ways – experience all part-part-whole combinations of the 'basic numbers' through many senses: in a visual, a tactile and a kinaesthetic way (sometimes also in an aural way, when they listed names of their fingers).

At that time they were willing to add and subtract within the number range 1-100. Then they drew schematic pictures, where tens instead of fives structured the numbers (Fig. 7). They could, for instance, draw 10-lines and unit-lines, and encircle the known part in subtraction, as in Fig. 7.

Finally, the teachers helped them to put those 'tens' together to a number line, marked at 10, 20 and so on, and later just at 50 and 100 (Fig. 7).

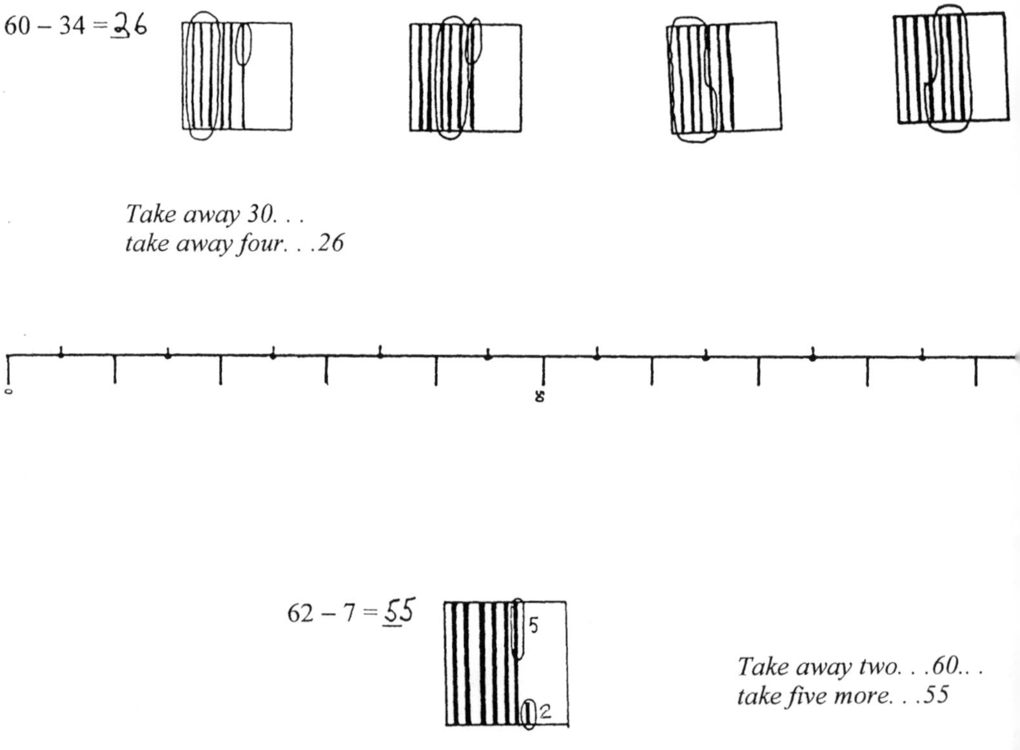

Figure 7

The 11 children expressing the earliest conceptions in the school beginner interviews, those with no correct answer on the subtraction problems, were interviewed again at the end of grade 2, as were the 13 children in the five control classes with comparable results in the beginner interview. Many kinds of problem were given. The research classes performed better in all contexts, but the most striking difference concerned the four answers on the mental calculation test within the number range 1–100. Eight of the 11 children in the research classes solved all these problems instantly and correctly, and all solved at least three of them. All of them gave satisfactory explanations to their answers, revealing how they could 'transform' one number combination to another, for instance 58 + 6 ... 'Just took away two from six (and added it on to 58) ... 64' – or how they split up the number at the 10-border – for instance 82 - 7 ... 'Took away two ... 80 ... five more ... 75'. All children gave direct answers, sometimes still explaining afterwards that they transformed, took away, added and so on, when they solved the nine questions within the number range 1–10 and the six within the number range 1–20. None counted, or used their fingers or any other concrete material.

In the control classes, only one of the thirteen children could solve all the four problems within the range 1–100 instantly and correctly. This child always used the idea of splitting up the number at the 10-border. A further five children in these classes solved all the problems correctly, but not through mental calculation. These children either used abacuses, notes, coins and so on, or the kind of counting and keeping track strategies observed among the pupils with maths difficulties. Nine of these thirteen pupils even solved many of the nine problems within the number range 1–10 through counting and keeping track, or through using concrete objects of some kind. Four of them estimated the answers several times, and one still answered several questions using 'limit names'.

4 CONCLUSION

Three conclusions can be drawn from this study. The first is that children's early intuitive ways of thinking about numbers and counting operations, and also their early misunderstanding of numbers and counting words, should be taken into consideration when mathematics is introduced.

The second is that *conceptual* knowledge of numbers and of counting operations – for instance knowledge of numbers as inseparably ordinal-cardinal or of subtraction as one single counting operation instead of as 'building up' and 'take away' – is developed in close mutual interplay with *procedural* development. What seems to be a good mental calculation strategy, and what also seems to be the kind of knowledge called 'known number facts', is often probably experiences of numbers that are grouped and structured in such a way that they become easily perceptible.

The third conclusion is that a way to relate conceptual and procedural knowledge (Silver, 1987) seems to be concrete problem-solving carried out through discussion between pupil-pupil and pupil-teacher, with ample time for individual reflections.

REFERENCES

Carpenter, T P and Moser, J M (1982). The development of addition and subtraction problem-solving skills. In Carpenter, T P, Moser, J M and Romberg, T A (ed), *Addition and Subtraction, a Cognitive Perspective*. Hillsdale: Lawrence Erlbaum.

Marton F (1981). Phenomenography – describing conceptions of the world around us. In *Instructional Science,* **10**, 177–200.

Miller, G A (1967). *The Psychology of Communication.* New York: Miller, Basic books.

Neuman, D (1987). *The Origin of Arithmetic Skills. A Phenomenographic Approach.* Acta Universitatis Gothoburgensis.

Neuman, D (1989a). *Räknefärdighetens rötter (The Roots of Numeracy).* Stockholm: Utbildningsförlaget.

Neuman, D (1989b). Early conceptions of subtraction. In Vergnaud, G (ed), *Actes de la 13e Conférence Internationale PME, Vol 3.* Paris.

Neuman, D (1992). The influence of numerical factors in solving simple subtraction problems. In Geeslin, W and Graham, K (ed), *Proceedings of the Sixteenth PME Conference, Vol 2.* Durham.

Piaget, J (1969). The Child's Conception of Number. London: Routledge & Kegan Paul.

Silver, E A (1987). Using conceptual and procedural knowledge: a focus on relationships. In Hiebert, J (ed), *Conceptual and Procedural Knowledge: The Case of Mathematics.* Hillsdale: Lawrence Erlbaum.

5

Modelling two-digit addition and subtraction with an empty number line

Koeno Gravemeijer
Freudenthal Institute, Utrecht University, Netherlands

SUMMARY

The traditional base-ten blocks used to teach two-digit addition and subtraction can be classified as a top-down approach. The prescriptions students are expected to follow are derived from the written algorithms, and the blocks themselves represent expert knowledge of our decimal system. In this paper an alternative approach is presented, where informal strategies of the students are taken as a basis for developing mathematical models. As such the empty number line is presented as a model that can be constituted in modelling solution strategies first, and functions as a model for mathematical reasoning later.

1 MANIPULATIVES

Traditionally, instruction on addition and subtraction algorithms is supported by the use of base-ten blocks like the Dienes blocks. In cognitive psychology these manipulatives are referred to as 'models'. These concrete 'models', visual 'models' and such are employed to embody mathematical concepts. The idea is to concretise abstract mathematical knowledge in order to make it accessible for the pupils. However, what a 'model' is in cognitive psychology differs from what a model is in the terminology of the modelling discussion. Blum (see ch. 1) among others, uses the term model to refer to a mathematical model. To underscore this distinction I will write 'model' to indicate usage in the cognitive psychology sense.

Within the framework of a teaching theory on realistic mathematics education – which is the framework of this article – it is useful to acknowledge a variation in level. These levels vary from informal mathematical models to formal mathematical models expressed in mathematical formula.

In cognitive psychology, Dienes blocks are thought to exemplify the base-ten position system. Furthermore the pupils will learn how to deal with this system by working with the blocks in a prescribed way. The characteristics of the positional system are incorporated in rules like

- always exchange groups of ten blocks for higher order blocks,
- notate the number of blocks in a strict order, corresponding with increasing values.

In general, the main concern however is not with the position system as such, but with the written algorithms. Therefore other rules are added, such as forcing the pupils to start with the smallest blocks, the ones, next move to the tens and so on. This is demanded because working from right to left is the standard procedure within written algorithms. Translated in terms of blocks, this means dealing with the blocks in order of their size.

We may infer that the source of any problems with the traditional approach lies in the forced character of this procedure. The directions which have to be followed are dictated by the written algorithm. From the standpoint of the student, however, those directions will be counter-intuitive.

Take for instance 85 – 59, starting with 85 represented with base-ten blocks.

Here it would only be natural to take away 5 tens first. Then the question follows, what is left if you take away nine ones? Knowing their basic facts the pupils might think of 15 – 9 = 6 and exchange one ten and five ones for six ones, or the pupils might realise that taking away nine is similar to taking away ten and adding one.

Solving the problem according to the standard procedure, one should start as follows. See if you can take away nine from the ones.

This cannot be done, so exchange one ten for ten ones. Then take away nine ones. Now check if you can take away five tens. Ok, take away five tens. Read off the answer: 26.

By the way, observe that the pupils does not have to realise that exchanging the one ten results in a total of fifteen ones, nor that $15 - 9 = 6$ is implied by taking away nine ones.

We may characterise this teaching approach as 'top-down'; the ready-made column algorithm and expert knowledge of the decimal system are taken as points of reference. The rods of ten, one-hundred squares and the one-thousand cubes are presented as ready-made models of units of a different rank. Furthermore the blocks are to be dealt with according to prescribed rules. This approach does not stay without consequences. Research reveals

- a lack of understanding,
- a lack of proficiency,
- a lack of applicability.

- Although pupils are expected to see what number of small unit-blocks the large blocks represent, this representation is not self-evident (Labinowitz, 1985). One of the reasons might be that the constitution of the groups of ten and higher is not of their own making (Cobb *et al*, 1992).
- Since the pupils do not understand what they are doing, it is hard to master the standard algorithms. Instead, the pupils develop 'repair algorithms' to make their results look right (Resnick and Omanson, 1987).
- To meet the problems with applicability, semantic structures of applied problems (change, combine/separate, compare, equalise) have been represented with visual 'models' to help the pupils discern addition or subtraction in applied situations. However this method did not solve the problem. Instead, research revealed the importance of domain-specific knowledge and informal strategies.

2 REALISTIC MATHEMATICS EDUCATION: COMMON-SENSE MATHEMATICS

Although manipulatives are thought to be concrete, the discussion above shows that they are not concrete in a meaningful way. To distinguish between these two interpretations, I will denote the latter interpretation of concrete with 'common', after common as in 'on common grounds' and 'common sense' (see also Freudenthal, 1991). In this sense working with manipulatives is not common but rather abstract; rules with an abstract justification prevent the pupils from common-sense reasoning. Therefore we may place these manipulatives on a level of rather abstract 'models', somewhere between familiar, common situations and formal mathematical knowledge.

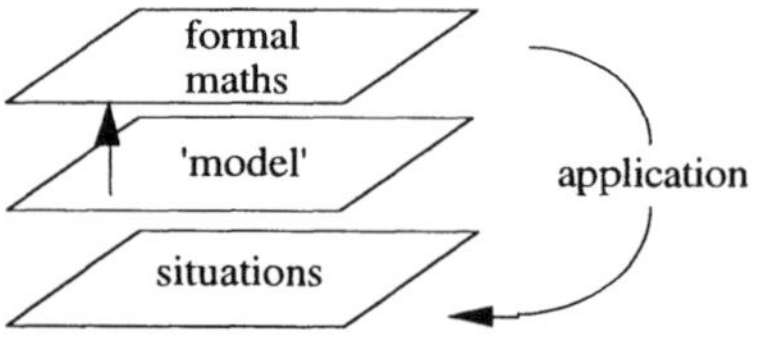

Looking at explanatory 'models' this way, it shows that these 'models' drift in the air; they are not connected with informal knowledge or anchored in familiar situations. The level of common situations is almost completely left out. Even common situations as applications are hardly incorporated, application is taken for granted. This is a consequence of the *top-down* approach, where the 'models' are derived from formal mathematical knowledge. In the following I will contrast this top-down approach with a *bottom-up* approach where the pupils create their own models by modelling informal strategies.

Freudenthal (1973, 1991) defines mathematics as an activity, within which mathematising plays a central role. He argues that in developing mathematics, formal mathematical knowledge constitutes the end point of a process of mathematising. In the top-down approach these end points are taken as starting points, an approach, which he would qualify as an anti-didactical inversion. Freudenthal's alternative is elaborated in what is called realistic mathematics education (Treffers, 1987; de Lange, 1987; Gravemeijer *et al*, 1990).

In realistic mathematics education, the informal strategies pupils develop when solving problems in common situations are taken as starting points. Next formal mathematical knowledge is construed in a process denoted as progressive mathematisation. This process may be described using the distinction Treffers (1987) makes between horizontal and vertical mathematising, the first referring to mathematising common problems, the second to the mathematisation of mathematical matter. Starting with solving problems in common situations, horizontal mathematisation produces models that are so close to reality that Blum (ch. 1) might prefer to call them real models. However, since they are the result of a mathematisation process, we would argue these are genuine mathematical models.

In line with Freudenthal's broad view on mathematising – which includes mathematising mathematical matter and subject matter from reality – we are willing to accept informal schemas and notations as mathematical models (see Gravemeijer *et al*, 1990 and Treffers, 1991). Anyway these models will become mathematical models for sure after a process of generalising and formalising. The informal models are then subject to vertical mathematisation, they become generalised and formalised and, in this way, a basis for formal mathematical knowledge is established. Later on, the same mathematical models will mediate in the opposite direction, facilitating the application of formal mathematics in other situations.

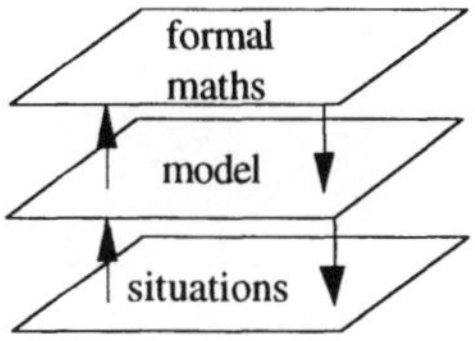

As we asserted above, the rule-governed operations with base-ten blocks do not correspond to a natural way of problem solving. If we compare these standardised procedures with the informal strategies pupils exhibit when solving applied problems the gap is even wider. This is shown, for instance, by an investigation involving 44 Dutch students (8-9 years olds), who had to solve the following problem (Vuurmans, 1991).

> A book has 64 pages. You have already read 37 pages.
> How many pages still have to be read?

The results show that there is a wide variety of solution procedures. More than half of the solution procedures can be categorised as a string method – whether adding on in steps (e.g 37 + 10 = 47, 47 + 10 = 57, 57 + 7 = 64, 27 added), or taking away in steps, (e.g 64 – 10 = 54, 54 – 10 = 44, 44 – 7 = 37). It shows that the string method is rather popular and fairly effective. However this method is not taught in schools, it is a strategy that children invent for themselves. This reveals another characteristic of the rule-governed manipulation with blocks – it surpasses already existing informal knowledge and strategies of pupils.

3 EMPTY NUMBER LINE

The string method is easily represented with jumps on an empty number line.

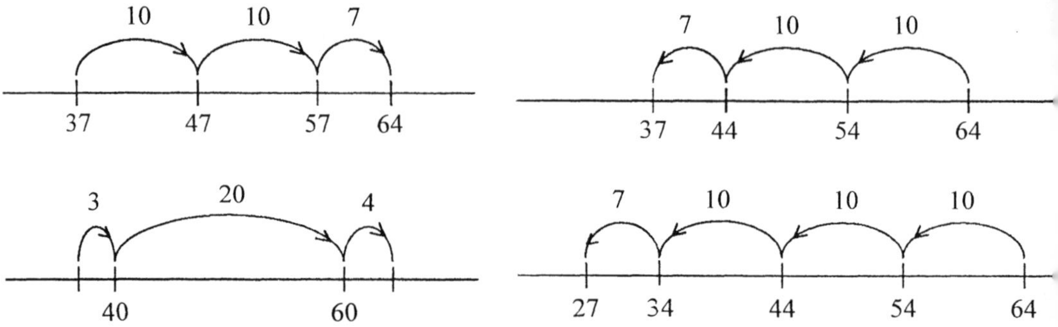

This empty number line also lends itself to skillful reckoning.

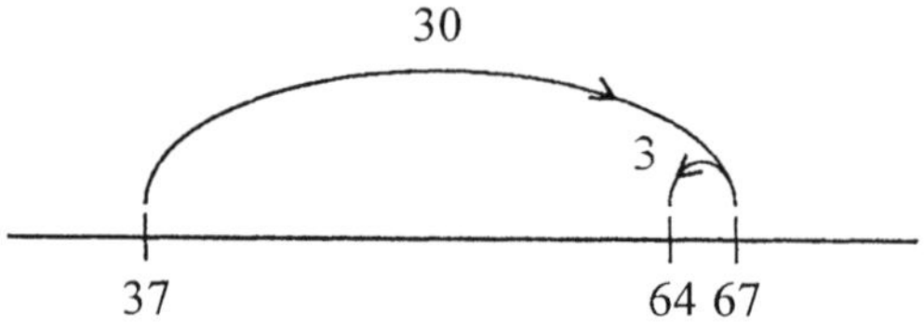

All-in-all there are good reasons to see that the empty number line can be used as an alternative for the aforementioned Dienes blocks. Or better, to conclude that, next to the set-type situations represented by the Dienes blocks, linear-type situations will have to be incorporated. It seems self evident to think of measuring and distances as initial situations for the development of the number line. Experiments with a teaching sequence, where the empty number line comes to the fore as a schematised version of a ruler or something similar, however, were not successful. This may have to do with the rigid character of the ruler. All distances have to be strictly proportional, which conflicts with the idea of placing unknown numbers on a number line. Pupils argued that you do not know where to put an unknown number – if you could indicate a certain place, you would implicitly specify its value. Moreover, taking the ruler as a base for the number line could result in an introduction of the number line as a visual 'model', instead of a mathematical model.

One can get around this problem by using a bead string to introduce the number line (Whitney, 1988; Treffers, 1991). First a bead string, with a hundred beads coloured in groups of ten, is used for counting games. The beads are counted from left to right, the number is indicated with a clothes pin that marks the total of beads to its left. Pupils find out that the decimal structure can be used to count efficiently. The pupils solve tasks like "Count 38 and add 24 more – which number do you get?" in a number-line-like way.

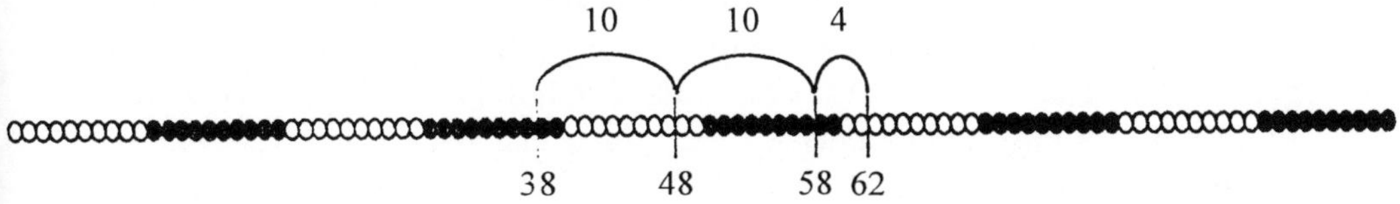

In this manner, counting games on the bead string anticipate the use of the number line – the empty number line is introduced to model the activities on the bead string. The bead string has several appealing qualities:

- it is isomorphic with the number sequence
- it fosters flexible use and curtailment
- it can be mathematised with an empty number line

We may note in passing that the isomorphism with the number sequence is an advantage over the base-ten blocks. The bead string represents the number sequence the pupils already know, base-ten blocks represent the decimal system which the pupil do not know yet, and which the blocks have to convey.

Note that the activities, the solution procedures, are modelled, and not the bead string as such. In hindsight we think that this is the main reason why this sequence gives rise to flexible use of the empty number line – since the pupils are not representing the bead string as such, there is no reason for them to worry about the exactness of distances. In the solution procedures the number sequence, not the distance, is the salient feature.

We can also put it another way – the empty number line has become a mathematical model. The bead string itself, however, has the character of a 'model' similar to the Dienes blocks, although the aforementioned differences should be acknowledged.

4 LEVEL RAISING

In solving problems the pupils mark the numbers they use on the empty number line. As a consequence, working with the number line functions as a way of scaffolding. This will facilitate the use of more advanced strategies, because symbolisation on the number line will help the pupils to keep track of where they are in their solution process. The next step in the course is to see if the pupils can model other common situations. This is one of the key principles in realistic mathematics education, to foster formalisation and level raising by generalising solution procedures and models to other situations. What is to be strived for is a shift in attention, from solving a given problem to a reflection upon the solution strategies from a mathematical point of view.

Compare, say, 75 – 67 and 75 – 8. In the first case, an interpretation of the subtraction as a *comparison* is the most effective approach, in the second case an inter-pretation as a *take-away* problem simplifies the calculation.

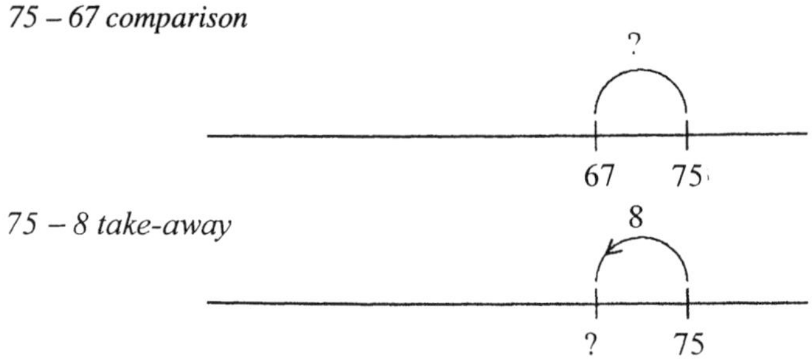

Similar shifts between 'counting on' and 'counting down' are observed with young children (Carpenter and Moser, 1984). The subtraction 9–7 for instance is solved by counting on: (7), 8, 9, → answer 2. The subtraction 9–3 is solved by counting down: (9), 8, 7, 6, → answer 6.

To be able to solve two-digit subtractions presented as bare sums in such a way pupils have to interpret the minus sign in this twofold way, as comparison and as taking away. In a teaching experiment (Veltman, in press) this twofold interpretation was related to 'taking away from the beginning' and 'taking away from the end'.

from the end

from the beginning

To many pupils it came as a surprise that both procedures lead to the same answer, although a short reflection made them realise that this makes sense.

In the same experiment the following task was presented in a written test.

Gary is baking a pizza. The pizza has to stay in the oven for 60 minutes.
Gary sets the cooker alarm on 60. When he looks again the hand on the alarm is at 34.
How long has the pizza been in the oven?

The results showed both interpretations.

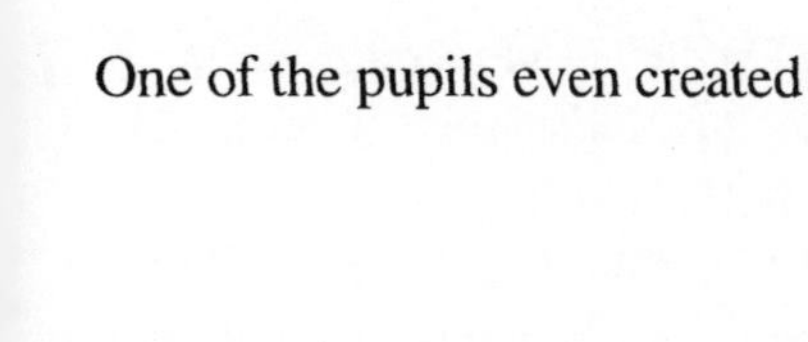

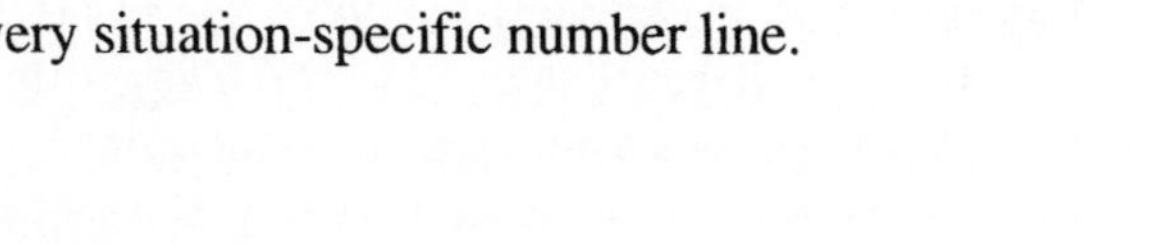

One of the pupils even created a very situation-specific number line.

When the shift towards conscious choices of solution procedures – based on the numbers involved – is made, the model gets a different status. Instead of a *model of* a situation, it may now function as a *model for* mathematical reasoning (see also Treffers, 1990; Streefland, 1985). The empty number line could be applied in this manner to justify a strategy for skillful reckoning, for instance

$$85 - 59 = 85 - 60 + 1$$

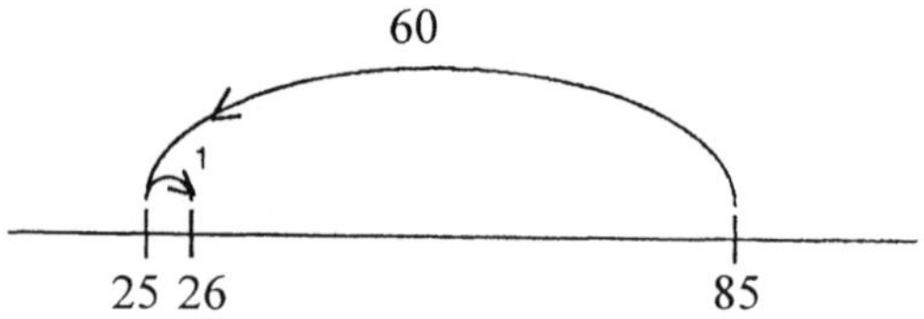

In this way realistic mathematics education constitutes a bottom-up approach, as an alternative for the traditional top-down didactics described in the beginning of this article.

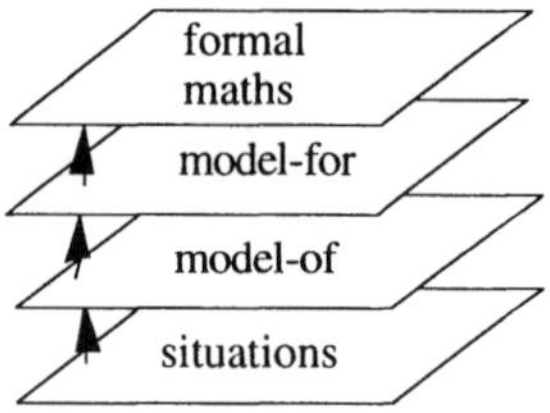

What is more, models so conceived will mediate in the opposite direction as well, fostering the applicability of formal mathematical knowledge. This also applies to informal mathematical knowledge – in fact, there is a constant interaction between mathematisation and application.

REFERENCES

Carpenter, T P and Moser, J M (1983). The acquisition of addition and subtraction concepts. In Lesh, R and Landau, M (ed), *The Acquisition of Mathematics Concepts and Processes*. New York: Academic Press, 7-44.

Cobb, P, Yackel, E and Wood, T (1992). A constructivist alternative to the representational view of mind in mathematics education. In *Journal for Research in Mathematics Education*, ***23***, 2-33.

Freudenthal, H (1973). *Mathematics as an Educational Task.* Dordrecht: Reidel.

Freudenthal, H (1991). *Revisiting Mathematics Education.* Dordrecht: Kluwer Academic Publishers.

Gravemeijer, K, van den Heuvel, M and Streefland, L (1990). *Contexts, Free Productions, Test and Geometry in Realistic Mathematics Education.* Utrecht: OW&OC.

Labinowitz, E (1985). *Learning from Children.* Amsterdam: Addison-Wesley.

de Lange, J (1987). *Mathematics, Insight and Meaning.* Utrecht: OW&OC.

Resnick, L B and Omanson, S F (1987). Learning to understand arithmetic. In Glaser, R (ed), *Advances in Instructional Psychology. Vol 3.* London: Lawrence Erlbaum, 41-97.

Streefland, L (1985). Wiskunde als activiteit en de realiteit als bron. In *Nieuwe Wiskrant,* **5**, 60-67.

Treffers, A (1987). *Three Dimensions. A Model of Goal and Theory Description in Mathematics Education: The Wiskobas Project.* Dordrecht: Reidel.

Treffers, A and de Moor, E (1990). *Proeve van een nationaal programma voor het reken-wiskundeonderwijs.* Deel 2. Tilburg: Zwijsen.

Treffers, A (1991). Didactical background of a mathematics program for primary education. In Streefland, L (ed), *Realistic Mathematics Education in Primary School.* Utrecht: CD-ß, 21-57.

Veltman, A (in press). *Rekenen tot 100* (master thesis).

Vuurmans, A C (1991). *Rekenen tot honderd.* 's Hertogenbosch: KPC.

Whitney, H (1988). *Mathematical Reasoning, Early grades.* Princeton.

6

Constructing a village: a cross-curricular activity

Panagiotis Athanasopoulos, Tasos Patronis, Despina Potari, Dimitrios Spanos, Vassiliki Spiliotopoulou
University of Patras, Greece

SUMMARY

In this paper a cross-curricular activity is discussed which allows primary school children to construct a real model of a familiar real-life situation. The project aims to investigate the rôle of children's personal experience, ideas conceptions and strategies in the modelling process. Moreover an attempt has been made to integrate various school subjects.

The whole experiment showed that children used mathematical or physical notions implicitly, while their values predetermined their actions on the whole modelling process. Moreover the integrated nature of the activity and the way it was carried out provided children with the opportunity to test and promote their mathematical conceptions and created positive attitudes in them.

1 INTRODUCTION

In recent years a lot of work has been done to incorporate real situations in mathematics instruction. A widespread belief on how this can be done, referred to as the 'utility argument' is that "mathematics instruction should enable students to practice applications and modelling in a variety of contexts where mathematics has instrumental services to offer without occupying in itself the focal point of interest" (Blum and Niss, 1991). Moreover some researchers emphasise that the use of mathematical models in social reality relies on a background of social values, intentions or assumptions (Keitel, 1989 and Skovsmose 1989, 1990).

Theoretically, it has been emphasised that an important stage in the modelling process is the construction of a real model of the original situation which "has to be simplified, idealised, structured, subjected to appropriate conditions and assumptions and made more precise by the 'problem solver' according to his/her interests" (Blum and Niss, 1991). In practice, however, this stage is often neglected. This is most important in the case of primary education, where young children learn meaningfully through social interaction and interaction with their own environment.

In this paper we discuss a cross-curricular project where children of a rural primary school (aged 8-9) were asked to design and construct a model of a village in which a team of thirty foreign children would spend their holidays. This activity can be characterised as a modelling activity in a broad sense, and allows children to construct a real model of a familiar real-life situation. It provides children with the opportunity to move from the real world to a real model by implicitly using ideas from mathematics and physics. Naturally, a stage of building a formal mathematical model is not expected from children of that age. The project itself enables children to act according to their values, intentions or assumptions, and their personal experience to be exploited and encouraged to develop.

2 THE RATIONALE OF THE PROJECT

The theme of the project was familiar to the children. The pupils involved live in a tourist village not very far from the sea. This allowed us to investigate the rôle of the children's personal experience in the modelling process.

In the building of a real model, decisions and choices have to be made which determine the actual construction. This means that children would decide which factors were important and which of them could be neglected. Thus children could pose some of the following questions.Which are the needs of the foreign children who are supposed to come and stay in the village, and how must the village be structured in order to meet these needs? How important is the aesthetic of the village? In addition children's ideas and conceptions on basic notions on mathematics and physics would be crucial during the process. For example, children's theories about the structure of matter could possibly influence the way that the mountains should be constructed.

The whole task is an attempt to integrate various subjects of the curriculum, meeting the children's own interests. This philosophy is not common in the Greek school curriculum. Recently two tendencies for such an integration have been encountered. First, environmental education has been given a priority in primary education, although it is still considered to be an extra-curricular area (Vasala, 1992). Second, some 'real situations' used in the mathematics classrooms are not in fact so 'real'. In particular, this holds for the mathematics textbooks, where either the problems are artificial or their solving has been already modelled. In contrast, the proposed cross-curricular activity involves a number of real situations in which children pose their own problems and develop their own strategies.

The openness of the activity allows children to develop their own strategies and plans, possibly conflicting with each other, according to their particular goals. The nature of the story, a social activity in itself, demands children to cooperate and make decisions together. Finally, children's ideas and conceptions could be revealed and promoted through the discussion on the problems that children had posed during the process (an example of such a discussion is desribed by Patronis, 1989).

3 THE PHASES OF THE PROJECT

The experiment was carried out in four phases. First, each child drew a plan of the village. In the second phase they worked in groups of four and made a more detailed plan. After a whole-class discussion on the group drawings (third phase), one group plan was chosen to be used for constructing the model (last phase). During all the phases, children were observed and interviewed; in the final phase children were also videotaped.

The designing phases offer us children's spontaneous views and representations of the space. Designing is also an essential part of the modelling process, since it involves "imagining nature without the 'unecessary' parts, and perhaps even emphasising some aspects more than others". In particular, in mathematics education we are interested in "the plan, the structure, the imagined shape, the perceived spatial relationship between object and purpose, the abstracted form and the abstracting process" (Bishop, 1988).

By making individual drawings, children become familiar with the activity itself. Moreover these drawings provide us with evidence on children's pre-existing beliefs and conceptions about the space. Additionally, in the second phase, children are given the opportunity to communicate and develop their ideas. A first analysis, both of individual and group drawings, provides us with specific questions to be answered in the later phases of the experiment.

The whole-class discussion helps children to express their ideas verbally and make them explicit. In this ways, children's strategies and conceptions emerge, while the debate makes their further development possible.

In the last phase the real model has been completed. Children's plans and strategies are now tested in practice, and sometimes changed. Here, we have an inverse situation to that of the designing phase – children pass from their own two-dimensional representations to the three-dimensional space. This inversion results from the need of realisation. The real model must look like the true object it represents – a village – and therefore it must be three-dimensional, otherwise it will remain only a plan. In this phase children were allowed to choose whatever they needed for the construction. This was in agreement with the 'openness' of the activity, and allowed children to construct the model they had designed.

4 THE OUTCOMES OF THE EXPERIMENT

The outcomes are discussed according to the following interrelated points.

- The modelling process
- The cross-curricular character of the activity
- Children's cooperation and communication of ideas
- The development of children's conceptions

The modelling process

Two main stages appeared in this process – the stage of designing and the stage of constructing the real model. The first one is more related to children's implicit mathematical ideas while, in the second, some matters of physics naturally emerged.

Children's decisions and choices during the designing stage were crucial, and determined to a certain extent the actual construction of the village.

Children have chosen several ways of representing the village in their drawings – globally, by designing a small part of it, or by showing only some representative elements (such as one house, one public building, one tree). In other cases the structure of the village reflected a functional choice – for example, a public building was separate from houses, sport centres, shops and suchlike.

Certain values underlying the children's choices have emerged from the group's presentation and defence of their work, and from the justification of their votes for which plan to be constructed. These can be categorised as aesthetic, functional and ethnological. Some were expressed verbally by the children while others, in spite of the fact that they were hidden, determined their actions, as is evident in the following extract.

- We drew a block of flats so children can live together.
- Roofs are not flat so that water falls down.
- In this plan, I like the curtains because of their colours.

The ecological environment greatly influenced the content and themes of children's drawing. The same holds for the social and cultural environment, which also had an effect on the choice of materials of construction and the instruments of drawing. An important part of the designing phase was the transformation from the real three-dimensional space to the two-dimensional drawing. A variety of ways of projecting the space were identified.

- Orthogonal projection, which is used to represent objects of relatively small height such as as lakes, squares, playgrounds.
- Representation of the facade of the buildings without any appearance of depth. This was often met in children's drawings, particularly in buildings' representation.
- Rotational projection.
- Drawing in perspective. In this case, children's techniques varied from primitive to more sophisticated ones. It is worth mentioning that different ways of projection were often used in the same drawing.

In making the real model children were confronted with the problem of transforming their plan into a three-dimensional construction. In the following dialogue, Kostas poses this problem for the whole village.

Kostas: Please tell me, will the village be standing up?
Int: What do you mean by 'standing up'?
Kostas: I mean will the plasticine stand like this...

Int: Could we make it differently?
Kostas: Yes as it is in the plan. (Meaning to make the model flat)

Children faced similar difficulties in the construction of specific elements of the village – as in making the boats, the mountains, the sun and the buildings. At first, strategies related to the two-dimensional space were used for the construction of these objects. However, as children faced the reality of the space where they live, they were led to move from two to three dimensions. In general, children did not consider the conservation of proportions passing from the plan to the space. Besides, children knew very well that they were constructing a model. It was not necessary for them to represent things exactly as they are in the real world.

The cross-curricular character of the activity

The nature of the activity led children to an interplay between concepts from various curricular areas. Children investigated questions related to these concepts in a real context, without being aware of their link to the formal curriculum. These are exemplified in the following diagram, which shows a road junction (Fig. 1).

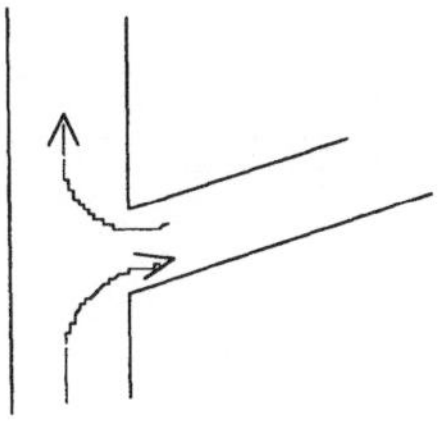

Figure 1

Junctions are considered more dangerous when they are not at right angles in the crossings, "it is easier to get in than to get out".

It seems that 'danger' comes from children's experience of the centrifugal force, and depends on the line's curvature. Angles are rather conceived as turnings, and in the case of a crossing we have a comparison between supplementary angles. The acute one was considered as the most 'difficult'.

In the construction of the mountains children considered different parameters – the structure of the matter, the static and the geometrical structure of the model – as well as ecological matters such as the existence of the trees and suchlike.

Some of the parameters that children took into account were not expected at the beginning. This shows the potential of such an activity to integrate a variety of areas like ecology, geography, astronomy, urban planning, and engineering.

Children's cooperation and communication of ideas

The openness and social nature of the activity, as well as the actual process of the experiment, involved children in cooperative work and interchange of ideas. This kind of interaction and interchange of ideas was especially apparent during the third phase of the project, in which children presented their work to the whole class.

The drawings from the other groups were questioned. This led the presenting group to justify their choices and strategies. Children, in this way, reflected on their ideas and made them explicit. This also enabled us to diagnose their conceptions. For example, when a group of girls presented their drawing, some children wondered if a certain house in perspective (Fig. 2a) was the correct one.

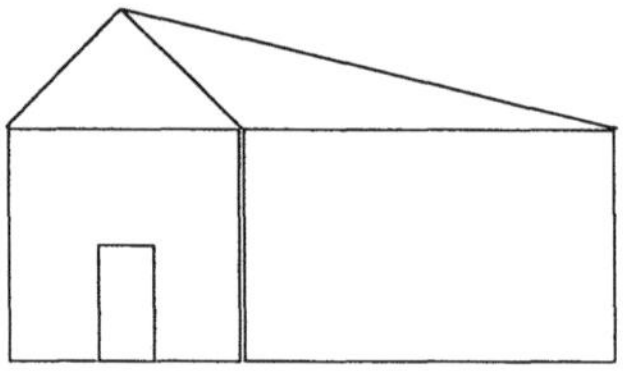

Figure 2a

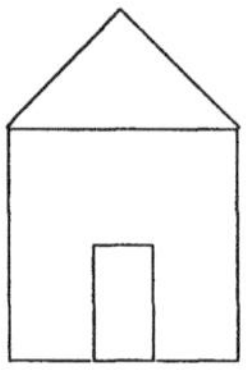

Figure 2b

Pupil 1: Why is the roof in this house drawn towards.....?
Pupil 2: We want to show that this roof continues at the back.
Researcher: Pointing to another house in the same drawing (Fig. 2b). What about this roof? Doesn't it continue at the back?
Pupil 3: We have drawn this house from another point of view. As it stands, the roof actually goes back.

The development of children's conceptions

Children's interaction and communication of ideas while working on the activity not only revealed their conceptions but also helped in their development. Vaso's story (below) indicates this development in the case of transformation from 2D to 3D. In constructing the boat, Vaso did not consider initially the third dimension. Her first attempt was a flat construction, which was finally rejected by the group as "not being able to stand up". Haris made a three-dimensional construction of the boat.

Int: Please tell me how did you make the boat? Where did you plan to put it?
Vaso: I wanted to put it down like this (flat on the sea).
Another child: But the sea is not as wide as needed for this.
Int: Haris has made a boat which stands. Which one is better? (Meaning the flat or the three-dimensional)
Vaso: Yes, but I've made it according to the plan.

Finally Vaso destroyed her own boat by accepting Haris' model.

Later, in the construction of the mountain, Vaso followed a similar strategy. Then, she placed her flat mountain vertically, but this was rejected by the group as inappropriate for planting trees. So, she modified her mountain by putting plasticine around it. In the building of the houses, however, Vaso no longer remains in her initial conception. She makes a house by joining four equal rectangular pieces of paper together – she wanted to make a cube.

The children, by using their ideas in a range of situations and having the opportunities to make them explicit, promoted their conceptions. In most cases this development took place through conceptual conflict.

5 CONCLUSION

In the light of our experience in this project, there is an implication for the contribution of similar activities in the school curriculum. One of the important perspectives of this experiment was the formation of a model of reality. In designing and constructing a model of their village, children were guided or misled by their conceptions and, at the same time, by their aesthetic or other values.

This process gave children the opportunity to use their implicit mathematical or physical notions that correspond to internal models or to the interpretation of reality. These notions influenced the whole modelling process, while the children's values predetermined action by suggesting choices and decisions. In this case, as it is also described by Cobb *et al* (1992), children participated in the formation of their own situations.

Moreover, this activity created positive attitudes in the children, who did not see any connection to their typical school subjects. What children learn in their normal classroom is mainly restricted to the particular context in which it has been performed. The project offered children the opportunity to use and develop ideas in a variety of situations. In such an integration, learning may become more meaningful and interesting for the children.

The process of building the model by working in groups had a direct influence on children's conceptions. Through debate, and argumentation of choices and strategies, children tested their ideas and promoted their conceptions. The project established an environment which encouraged cooperation and reflection on decisions and actions. Moreover, the activity of engaging children in active participation promises a change in the teacher's rôle. It limits teacher interference and teaching guidance, and helps both teachers and pupils to change their attitudes. This aspect needs further investigation.

Acknowledgments

The authors wish to acknowledge the valuable assistance given by A Panoutsopoulos and M Maniati.

REFERENCES

Bishop, A (1988). *Mathematical Enculturation.* Dordrecht: Kluwer Academic Publishers.

Blum, W and Niss, M (1991). Applied mathematical problem solving, modelling, applications and links to other subjects - state, trends and issues in mathematics instruction. In *Educational Studies in Mathematics,* **22**, 37-68.

Cobb, P, Yackel, E and Wood T (1992). Interaction and learning in mathematics classroom situations. In *Educational Studies in Mathematics*, **23**, 99-122.

Keitel, C (1989). Mathematics education and technology. In *For the Learning of Mathematics,* **9**, 7-13.

Patronis, T (1989). Interaction by open discussion and "scientific debate" in a class of 12-years old pupils. *Proceedings of the 13th PME Conference.* Vol 3. Paris.

Skovsmose, O (1989). Models and reflective knowledge. In *Zentralblatt für Didaktik der Mathematik,* **21**, 3-8.

Skovsmose, O (1990). Mathematical education and democracy. In *Educational Studies in Mathematics,* **21**, 109-128.

Vasala, P (1992). A presentation of environmental education (in Greek). In *Education Contemporaine*, **67**.

7

Modelling and applications of mathematics in the primary curriculum

Max Bell
University of Chicago, USA

SUMMARY

At least for primary schooling in mathematics, the numerous reform movements over many decades recommending substantial attention to uses of mathematics have had nearly zero effect. The 'basics' of arithmetic essentially always win out over modelling the uses of arithmetic. The paper describes a few of the compelling reasons why skills and basics have so consistently crowded out uses and mathematical modelling over those many decades, and argues that we better address those or we will lose again.

The paper describes as one main problem that there are very clear specifications of objectives, teaching procedures, testing, and reporting for basics, but no competing clarity in specifications for modelling and uses. We can do something about that, and we should, even over the short run. What we have given teachers to work with – the day to day curriculum expressed in basal books – gives them no help except for teaching basics, and even that at a glacially slow pace. We can do something about that, and most primary teachers can implement the results quite effectively.

1 A WORK OF CURRICULUM DEVELOPMENT

The primary curriculum of the University Chicago School Mathematics Project (UCSMP)

The elementary materials component of this developmental project is committed to building an example – an existence proof – of a primary curriculum with such features as the following.

- It should be mathematically much richer than the present norm in the United States.
- It should build on, rather than ignore, the substantial capabilities of young children that are revealed by research studies.
- It should build concepts and skills in cycles of recurring and expanding emphasis, with informal initial exposures which come normally a couple of years before mastery is seriously expected.
- It should be characterised by partner and small group collaborations, together with teacher input, as children build concepts for themselves.
- It should be usable by most teachers, with most children, in a broad range of different schools, culture, and socioeconomic settings.
- It should attend at every point to the applications of mathematics in a mathematical modelling framework, and be linked in natural ways to other school content such as science, language, arts and suchlike.

The curriculum example, *Everyday Mathematics,* has been completed for grade K-3 with funding from various sources (see Bell *et al*, 1987-92). Development of the curriculum for grades 4-6 is now beginning, with funding from the National Science Foundation.

Making a curriculum that emphasises modelling in primary schools

My focus in this paper is on primary schooling. My concerns here will be mainly the content and pedagogy of primary school curricula – *what* to do, and *how* to do it – and will only touch on the difficult problems of assessment of results and of more adequate preparation of teachers.

The basic arguments for my concern go as follows.

- Reform recommendations, over many decades, that advocate emphasis on mathematical modelling and 'teaching mathematics so as to be useful' have had little effect on what is done in schools, especially in primary schools.

- One main reason for the minimal effects is that specifications for teaching and testing basics are very clear and very compelling, but our competing recommendations for new emphasis seldom include comparably clear specifications. For example, 'fuzzy prayers' to include problem solving in first grade are unlikely to compete with teaching of pure skills such as 'teach addition with regrouping of two numbers less than 100'.

- I believe we can provide clear competing specifications for curriculum emphasis beyond the basics. Included here, as examples, are some outlines of what most people should know in addition to 'basics', of 'application classes' for arithmetic, and of mathematical modelling processes. For each of those outlines, fulsome amplification exists elsewhere.

- It is possible to turn such specifications into widely-usable curriculum materials and good advice for teachers. A few examples of our own attempts as an existence proof for that proposition are included.

2 REHABILITATING MODELLING AND THE USES OF MATHEMATICS IN PRIMARY SCHOOLING

Advocacy of linking mathematics to its uses goes back a long way

In nearly all of the proposed reforms of mathematics education advocated in this century, appeals are made for linking mathematics to its uses, and for making problem solving a main emphasis in school mathematics. Some typical examples are quoted in the Appendix. In the United States, the National Council of Teachers of Mathematics *Standards* continue that tradition of advocacy. However, at least in this country, all these recommendations have had few durable effects on the school experience of most children.

Easily-specified 'basics' always crowd out poorly-specified 'problem solving'

The rhetoric about the importance of problem solving and 'teaching mathematics so as to be useful' has been matched over the years with confusion and naivety about what problem solving is and how to teach it. As a result, those emphases have lost out to skills, drills, and symbol pushing in arithmetic and algebra. Perhaps this is so because arithmetic skills are relatively easy to formulate as well-structured curriculum and remediation schemes. Arithmetic skills are also easy to state as behavioral objectives, which can then be sorted into pseudo-precise grade-by-grade curriculum specifications. With such specifications, it is easy to propose measures of 'mastery', and to assign scores to children, to teachers, and to schools – the results of which are sometimes put to alarming uses in media headlines.

Hence, the arithmetic and algebra skills that are relatively easy to specify, organise, and assess, become the school curriculum. Problem solving, when attended to at all, tends to be embodied as 'word problems', whose main purpose is to exercise the arithmetical skills that, by default, have become the focus of the curriculum.

Things needed to change that state of affairs can be specified

To change that state of affairs in primary schooling at least five things are needed.

1 Clear specifications of what most people (not merely specialists) need to be comfortable with by the time they leave school, in order to use mathematics effectively and confidently throughout their lives.

2 Clear and simple specifications, and then teaching, of effective processes and heuristics for problem-solving processes. In other words, a lot of experience with the processes of mathematical modelling.

3 Outlines that categorise the main applications of arithmetic in primary school mathematics, as guides to curriculum coverage, and experience with number models and handling data.

4 A considerable amount of interesting data, problems, and project material along with the determination to make those central in school curricula.

5 Specifications of new pedagogical methodologies to make modelling in arithmetic a practical classroom emphasis that genuinely engages the energy and creativity of children. These must be set up in cooperation with the teachers.

The first four of these are discussed briefly in the section that follows. Some examples of the fifth – effective pedagogical technologies – are, we hope, exemplified in the *Everyday Mathematics* curriculum itself.

3 SPECIFICATIONS

3.1 What most people need to know about mathematics to be effective users

The current consensus of the mathematics education community on what should be taught is embodied in the curriculum and evaluation standards of the National Council of Teachers of Mathematics (NCTM, 1989), probably more effectively than in any past time in the United States. For many years we have been guided in curriculum efforts, in ways that have a modelling bias, by various iterations of the above list (an early version of which is given in Bell, 1979). As it happens, our list is consistent with the current consensus – even when considering its imperfections, we find that a one-page specification has some advantages over those containing hundreds of pages.

3.2 Achieving greater clarity about modelling or problem solving processes

We believe that clarity about problem-solving processes is best achieved by adapting the well-known conventions and heuristics of mathematical modelling. These conventions and heuristics can then be built into mathematics instruction at all levels, including primary mathematics. As an overall guide, we have found the following summary statement helpful (see Synge, 1961).

> The use of applied mathematics in relation to a physical problem involves three stages.
> 1 A dive from the world of reality into the world of mathematics.
> 2 A swim in the world of mathematics.
> 3 A climb from the world of mathematics back into the world of reality, carrying a prediction in our teeth.

However, more detail than that is needed as a guide to curriculum building. For that purpose we use the *process diagram* shown in Fig. 1, which is distilled from conversations with many people who use mathematics on a professional basis. These people say that, in modelling situations, they may visit any or all of the nodes in this diagram, in no particular order, and often with considerable bouncing around and recycling – one may go anywhere in the diagram from anywhere else.

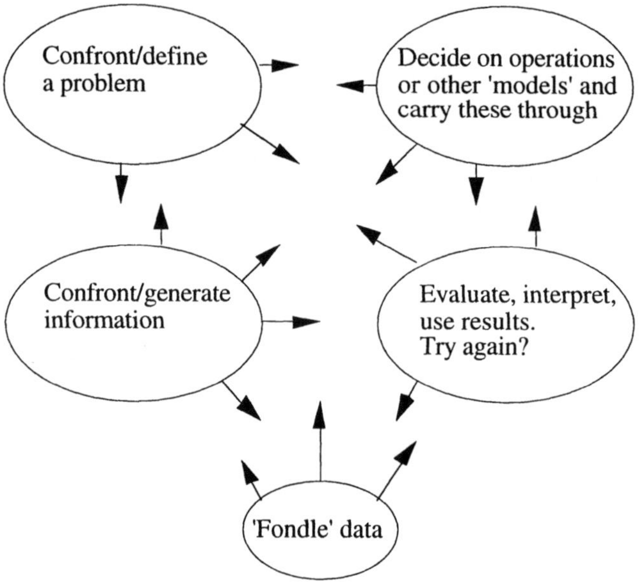

Figure 1. Processes of mathematical modelling

We note that the mathematical structures and procedural skills that dominate present US curricula, at all levels, are embedded in a single node of this diagram. That node is important and cannot be neglected, but it is just one piece of the modelling puzzle.

Even simple word problems would usually involve four of these nodes, emphasising confronting (rather than defining) a problem and using (rather than generating) data. In our programmes we ask children, from kindergarten on, to pose problems or tell number stories and generate data.

We believe that experience with each node of the diagram can be systematically built into the curriculum, but sometimes untidy navigation among the nodes requires judgments and intuitions that are difficult to teach in specified lessons. This is generally true of heuristics, as opposed to algorithmic processes, and can probably only be dealt with by providing for considerable modelling experience over a number of years.

The following outline seeks to show that the countless applications of arithmetic can be sorted into a relatively few teachable categories, as one basis for curriculum building. It is a substantially adapted version of a much longer treatment (see Usiskin and Bell, 1980) and it is organised around the diagram of components of the mathematical modelling process shown in Fig. 1 above.

Application classes for numbers and operations of arithmetic

1 *Confronting or Defining a Problem*
This may either precede, follow, or be mixed in with any of the activities that follow.

2 *Confronting or generating numerical information*
Most of the billions of numbers used daily in the common world fall into one of the following 'use classes' – often such numbers are components of n-tuples or data sets.

Counts	Ratios (scalars)
Measures	Estimates of any of the above
Locations in reference frames	Codes or identification numbers, often alphanumeric

3 *Fondling data*
Of the many things people do to bridge the gap between having numbers and deciding what to do with them, one or more of the following things tend to be done.

Change notation	Transform a data set
Use approximate values	Display a data set

4 *Deciding on operations or relations for arithmetic models*
Most of the countless daily uses of operations can be sorted into just a few application classes. Whether, for example, a 'change' situation calls for addition or subtraction depends on what is known or missing. Similarly for the multiplication and division application classes.

Addition & Subtraction	Multiplication & Division	Powering
Parts and total	Forming or using rates	Growth and decay
Change	Acting across	Size dimensions
Compare	Scaling and ratios	Notations

5 *Evaluation, interpretation, and recycling of 'answers' from models*
This revisits the other nodes of the modelling process, and may either be an endpoint or a new beginning.

3.3 Conceptual structures for sorting out the applications of arithmetic

To be competitive with the teaching of the basics of arithmetic computation, we need clear and compelling conceptual structures to guide teaching, focused on the *uses* of arithmetic. I believe that a start on that was made in the three-part book *Applying Arithmetic* (Usiskin and Bell, 1983), which elaborates on the summary of application classes given above.

The modified application classes of our K-3 curriculum differ from the ones in *Applying Arithmetic* and the UCSMP secondary school textbooks mainly in the merging of uses of addition and subtraction, on the one hand, and multiplication and division on the other hand, resulting in a much smaller number of characteristic application classes. It is important to note that different people think of the same problem in different ways. For example, number models for problems involving odometer readings and elapsed distances are legitimately thought of by some as 'change', by others as 'comparison', and by still others as 'parts and total'.

Some functions of the application class structure include giving *developers* guides to the inclusion of a broad range of problems in curriculum building, giving *children* a variety of ways to think about how to choose number models, and persuading *teachers* that teaching the uses of arithmetic is a finite problem, with a structure that can be as clear as that for teaching the basics of computational skills.

The point here, of course, is not to defend either the original application classes of *Applying Arithmetic* or the modifications in our K-3 curriculum, but merely to argue that simple structures for arithmetical modelling are achievable, and that they can be competitive with skill and drill structures.

3.4 Providing interesting data, problems, projects, and exploration materials

Many materials are needed with genuine links to common life and to various aspects of the natural world (including geography), the man-made world, and the social world (including sports and recreations). Fortunately, in contrast to a decade ago, there is a wealth of such materials under development or in recently-published books. For example, *all* of the curriculum projects currently funded by our National Science Foundation, are dedicated to producing more such material and to making modelling a central focus. I find this uniformity of purpose quite surprising, and we should note it well. Provisions for genuine applications and modelling, that have been rare in schools, have become mainstream in the efforts to reform school mathematics, which may have its own hazards.

REFERENCES

Begle, E G (1979). *Critical Variables in Mathematics Education.* Washington: The Mathematical Association of America.

Bell, M S *et al* (1987-1992). *Everyday Mathematics, Grades K, 1, 2, and 3.* Evanston: Everyday Learning Corporation.

Bell, M S (1979). Teaching mathematics as a tool for problem solving. In *Prospects*, **9**, 311-320.

Freudenthal, H (1968). Why to teach mathematics so as to be useful. In *Educational Studies in Mathematics*, **1**, 3-8.

National Council of Teachers of Mathematics (1989). *Curriculum and Evaluation Standards for School Mathematics.* Reston: National Council of Teachers of Mathematics.

Synge, J (1961). The remark is quoted by M R Kenner in Mathematical Education Notes. In *American Mathematical Monthly,* **68**, 799.

Thorndike, E S *et al* (1923). *The Psychology of Algebra.* New York: Macmillan.

Usiskin, Z and Bell, M S (1983). *Applying Arithmetic: A Handbook of Applications of Arithmetic.* ERIC document nos. ED 264 087, ED264 088, ED 264 089.

APPENDIX

Samples of good advice on teaching uses of mathematics

1

E H Moore, in his 1902 American Mathematical Society presidential address, said that schools should abolish 'watertight compartments' in which algebra, geometry and physics are taught. They should stress interrelationships of mathematics with science. Moore cited approvingly the work of a contemporary, John Perry, who urged emphasising interrelationships of mathematics with science, making mathematics concrete by use of laboratory-teaching techniques and applications, and bringing into lower grades many simple ideas usually postponed to later grades.

2

"Problem solving in school is for the sake of solving problems in life. Other things being equal, problems where the situation is real are better than problems where it is described in words. Other things being equal, problems which might really occur in a sane and reasonable life are better than bogus problems and mere puzzles." (Thorndike, 1923)

3

"Since mathematics has proved indispensable for the understanding and the technological control not only of the physical world but also of the social structure, we can no longer keep silent about teaching mathematics so as to be useful. In educational philosophies of the past, mathematics often figures as the paragon of a disinterested science. No doubt it still is, but we can no longer afford to stress this point if this keeps our attention off the widespread use of mathematics and the fact that mathematics is needed not by a few people but virtually by everybody." (Freudenthal, 1968)

4

"Mathematics is a lot of fun for a small number of individuals. For even a smaller number mathematics provides a profound esthetic experience. If that were the whole story it would not be possible to justify the emphasis given to mathematics in our school programs. The real justification for teaching mathematics in our schools is that it is a useful subject and in particular that it helps in solving many kinds of problems." (Begle, 1979)

8

Statistical simulations

Mary Rouncefield
Chester College, UK

SUMMARY

In statistical simulations probability and statistics appear as interwoven strands of the same theme, rather than as disjoint areas of mathematics. Real data can be used to estimate the probabilities required for the simulation, and the pupil can apply mathematics in a real context.

This paper outlines some simple statistical simulations for younger pupils, or for older students meeting simulations for the first time. Details of the simulations are given, to enable the teacher to use them as examples in the classroom. Guidelines are also provided for pupils and teachers wishing to design their own simulations.

1 WHAT IS A STATISTICAL SIMULATION?

A statistical simulation is a model of a real situation in which chance or random factors are involved. In a queuing situation, for example, a simulation can be used to help answer questions such as the following.

- How many pumps are needed at a petrol station?
- How many pay booths are required at a toll bridge?
- Is a central queuing system more efficient than queues at separate counters in a bank or post office?
- Will the introduction of an appointments system mean that a doctor's patients will have to wait shorter lengths of time for consultation?

In order to answer these questions the simulation will need to model both the arrivals of cars (or customers or patients) and the time taken to 'process' each one. Chance elements come into play at both ends of the queue, arrivals and departures, and these chance factors must be built into the model. This is a fairly advanced statistical simulation, and is explained in Rouncefield and Holmes (1989). The purpose of this paper is to show how simple statistical simulations can be a valuable classroom activity for younger pupils, and how they can be extended and developed to model fairly complex situations such as queues.

Modelling random processes

For any situation involving an element of unpredictability or randomness, some mechanism for randomisation is required in the simulation. This mechanism can range from a physical object such as a die, to an 'abstract' random-number generator in a computer. A physical object which the pupil can see, touch and understand is ideal for the younger student, and for an older pupil's first experience in statistical simulations. Examples of randomisation mechanisms include the following.

- Tossing a coin
- Throwing a die
- Rotating a spinner
- Making a selection from numbered raffle tickets or discs
- Random-number tables
- Calculator or computer random-number generators

2 EXAMPLES OF SIMPLE SIMULATIONS

This section describes two simple simulations suitable for younger pupils, although the first can be used as a starting point for older students.

Collecting free gifts or stickers

'Batman' stickers are given away as free gifts with a certain brand of breakfast cereal. The problem is to decide how many packets of cereal must be bought in order to collect a complete set of 6 stickers.

Actually buying the packets of cereal is an expensive and lengthy process, so a simulation can help. Each sticker can be allocated a number between 1 and 6, and a die used to model the chance factors in the situation.

Sticker	Score on die	Tally
Batman	1	
Cat Woman	2	
Penguin	3	
Batgirl	4	
Riddler	5	
Black Widow	6	

Each pupil in the class throws a die and records the result of each throw, continuing until at least one of each of the gifts on the list has been obtained. The number of throws of the die can then be counted, giving the number of packets of cereal which have to be purchased.

At this stage, results can be collected from all the pupils in the class, compared and discussed. In my experience, there is usually a wide range of results (58 being the highest so far). This can generate a great deal of discussion.

If the teacher reiterates the original problem "How many packets of cereal do I need to buy in order to collect a complete set of free gifts?" pupils will give a variety of answers.

– Well ... at least six
– Up to 30
– More than that – I needed 37!

If the teacher insists on a single number as the answer, the mode is usually suggested first. Eventually a pupil will propose finding the mean – hopefully! If the class results have been recorded as a frequency table, more discussion will ensue as to how the mean is to be calculated.

- Is this a good model?
- What assumptions are made in this simulation?

As pupils are likely to have met this situation in real use, they may be able to discuss the validity of this model in some detail. One class of pupils worked on this simulation, while at the same time there was a free-gift promotion with a popular brand of breakfast cereal. They decided to adapt their simulation to model collecting 12 gifts, and wrote to the manufacturer about the conclusions.

- None of the pupils, or their friends, had actually managed to collect a full set of 12 gifts before the promotion ended.
- Their simulation showed that it was indeed very unlikely that anyone would be able to complete their collection in such a short time.

Here are two assumptions made in the simulation.

- Free gifts are made and distributed in equal numbers.
- They are distributed evenly, and at random, throughout the retail system.

Any deviations from these assumptions may be difficult to investigate or to build into the model. If the manufacturer has deliberately made a smaller quantity of one of the stickers, how could the model be changed? Pupils may be able to consider changing to unequal probabilities if they can use spinners, random numbers or dice with more than six faces. These are discussed in later examples.

Smartie simulation

Smarties are coloured sweets, with chocolate inside. There are 8 colours.

Problem: Are the colours in a tube of Smarties fairly distributed?

Pupils need a real tube of Smarties for this investigation, so this is always a popular activity. Firstly, the Smarties need to be arranged in lines of the same colour, making a bar graph. Some pupils will find that they have an uneven distribution of colours, perhaps even with one colour missing entirely. Is this 'fair'? How can a 'fair' tube of smarties be simulated? As there are 8 colours a six-sided die cannot be used – although an octahedral die could be.

An alternative is to write each of the numbers 1 to 8 on a piece of paper, and to place these in a large box. After allocating a number to each colour, a volunteer can be asked to draw a number out of the box, as in a raffle. The number can be recorded, and then replaced in the box for the next draw. This process will need to be repeated 40 times or so, in order to simulate a tube of Smartie colours. As this will take a long time, pupils are usually quite happy to use a quicker method. Single-digit random numbers can be introduced quite naturally at this stage.

After examining a page of these, and discussing how they were generated, many pupils will be happy to use their own calculator – or the classroom computer – to generate random numbers.

Each pupil must generate sufficient random numbers to represent a tube of Smarties – the same quantity as the number of sweets in their own real tube of Smarties. Among the simulated tubes of Smarties, there will probably be some which have a colour missing, or other uneven distributions. While the distribution in each individual pupil's simulated tube will be different from that in their real tube, a class display of bar graphs showing real and simulated results will reveal more similarities than differences.

At this stage pupils have acquired an idea of what a simulation is, and may be ready to design their own simulations.

3 STEPS INVOLVED IN SETTING UP A SIMULATION

Outlined here is a list of steps describing the process of setting up a simulation. Younger pupils, with no knowledge of statistical distributions, may simply estimate probabilities using relative frequencies obtained from observations. Thus steps 3, 4 and 5 need only be considered by older students.

Setting up a simulation

1 Observe the real situation and identify the important variables. For example, these may be the birth and death rates of a threatened species, or the time intervals between the arrivals of patients at a hospital casualty department.

2 Decide on the main processes in the situation. A flow diagram may be helpful. Which are chance processes needing estimation of probabilities?

3 Can any of the features be modelled by a mathematical or statistical distribution?

4 What are the assumptions made by these mathematical models? For example, can we assume that casualty patients arrive independently of each other? This assumption would not be true of road accident victims. Check all your assumptions against reality. Are they close enough to the reality?

5 What parameters do you need to know? Will you need to estimate the mean and variance of doctors' consultation times, say, be taking a sample of observations?

6 Take a sample of observations in order to estimate the values of the parameters or probabilities.

7 Set up your simulation and run it several times.

8 If possible, compare your results with reality.

9 If your simulation is not producing realistic results try to incorporate additional factors which may be influencing the situation, or try to replace a simplistic relationship between variables with a more sophisticated one.

10 If your simulation is realistic, use it to predict the outcome of changing a feature in the real situation.

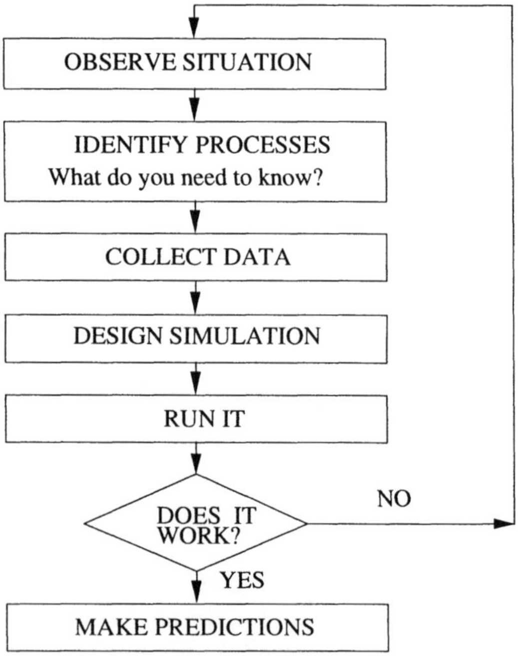

Figure 1

4 MODELLING TRAFFIC FLOWS

One-way street simulation

The simplest traffic simulation can be based on a one-way street. How long do the pedestrians have to wait before crossing the street? Is there a case for a 'pelican' crossing with traffic lights. One possible experiment is outlined in Fig. 2, taken from Hudson and Rouncefield (1991).

Collecting data

Kim and Robert decide to carry out a survey.
Robert times Kim crossing the street.
It takes her 20 seconds to walk across safely (using the Green Cross Code.)
It is a wide one-way street with traffic moving very quickly.

Robert has a watch with a second hand.
He says 'Now' every 20 seconds.
Kim has a recording sheet.
Each time Robert says 'Now', she writes how many vehicles have passed in the last 20-second period.

Recording sheet - James Street (one-way) - number of cars in 20 second time periods, 18 June

Time period	1	2	3	4	5	6	7	8	9	10	11	12
Number of vehicles	2	2	0	0	0	1						

Time periods	13	14	15	16	17	18	19	20	21	22	23	24
Number of vehicles												

Figure 2

Pupils can use the data provided, or they can collect their own data on a one-way street near their school. First of all, the results can be examined to find the longest time a pedestrian would have to wait to cross the road. In order to get a fuller picture more data is needed, and this will take a considerable time to collect. So the data already collected can be used to generate further results in a simulation.

A frequency table can be drawn up, and relative frequencies calculated. These relative frequencies can be used as probabilities, and further results generated using random numbers or a spinner, as in Fig. 3.

Relative frequencies

We can use relative frequencies to estimate the chance or probability of each result

Number of vehicles	Frequency	Relative frequency
0	29	$\frac{29}{60}$ 0.5
1	14	
2	12	
3	5	
Total	60	

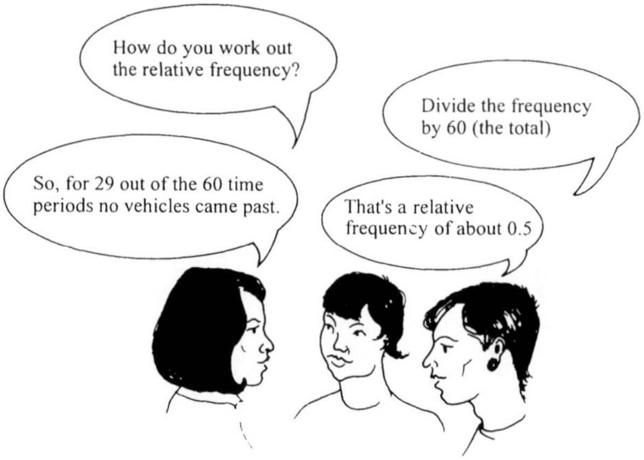

(1) Copy the table and fill in the other relative frequencies.

Give your answers correc: to 1 decimal place

■ Now make a 'One-way street' spinner to do the simulation experiments

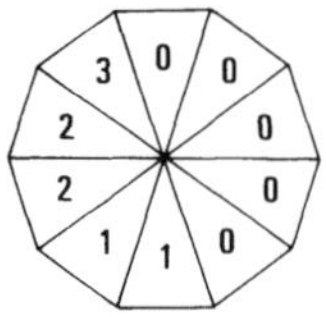

Try to explain why it is designed this way

Figure 3

The class can then discuss the original problem "Is a pedestrian crossing needed?" A simulation can be run to see what would happen if a pedestrian crossing were introduced.

- How long will pedestrians have to wait (on average)?
- How long will vehicles have to wait (on average)?

This one-way street simulation has some important features.

- Real data are used.
- Relative frequency is used as an estimate for probability.
- Statistical and probability ideas and skills are linked, and used in conjunction. In many cases, pupils work either with probability or statistics, rarely with both. In the minds of many pupils and teachers statistics and probability are separate entities, and links between them rarely made.

Two-way street simulation

Using the skills and ideas acquired in the one-way street simulation, pupils can progress to a more complex simulation – a two-way street.

- Traffic approaches from two directions. Two sets of observations are required to generate probabilities so that a two-way flow can be simulated.
- Pedestrians arrive at both sides of the street, wishing to cross over. Again two sets of data are needed to simulate this two-way flow.

Altogether flows are involved in the simulation, a team of pupils can work together using a map and toy cars and figures as shown in Fig. 4.

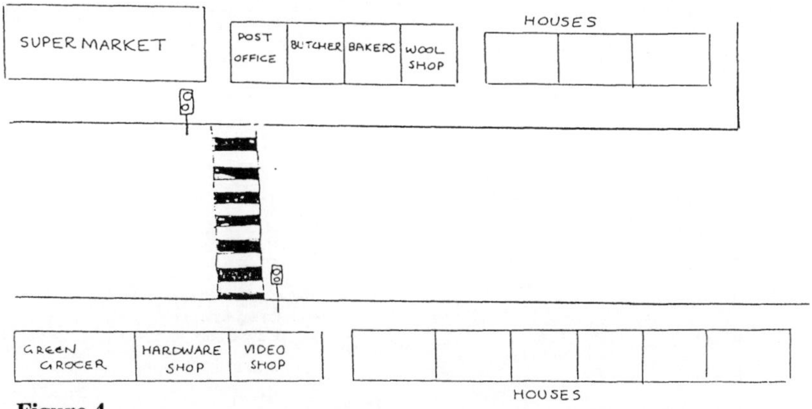

Figure 4

The simulation can be run both without and then with the pedestrian crossing. A more enjoyable, although possibly more chaotic, version of this simulation, can be run by drawing a roadway on the playground in chalk. Pupils themselves can take the rôles of pedestrians and cars.

5 POPULATION MODELS

Many pupils are interested in ecology and conservation issues, they can use the library to research data on the breeding and mortality of endangered species.

The breeding cycle of the royal albatross is suited to a simple simulation, as birds produce only a single egg once every two years. Suppose there exists only one small colony on a mainland inhabited by humans. Apart from this one group of birds living on a peninsula at the southern tip of South Island, New Zealand, all other breeding colonies are found on uninhabited islands. We wish to model the survival and growth of this colony.

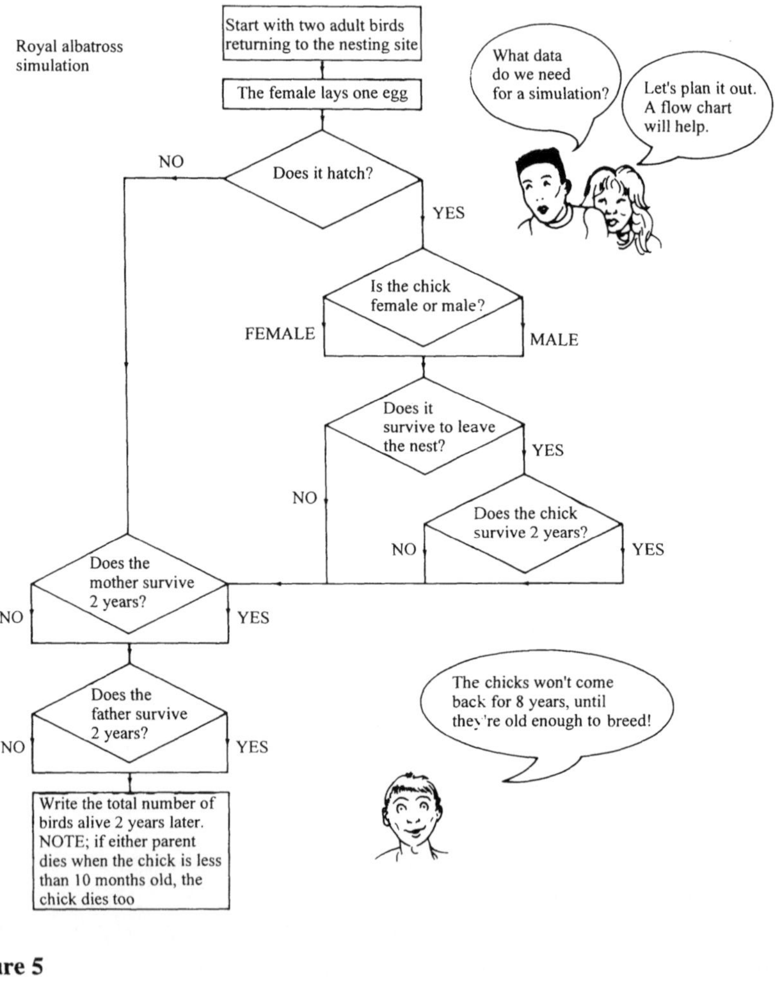

Figure 5

Fig. 5 shows a flow chart showing the processes involved in modelling a two-year breeding cycle for a pair of royal albatross (from Hudson and Rouncefield, 1992).

Albatross simulation

Stage 0	Two adult birds return to nesting site. The female lays one egg.
Stage 1	*Does the egg hatch?* Pick a random number: 1,2,3,4,5,6 or 7 = YES 8,9, or 0 = NO # Go to stage 5
Stage 2	*Is the chick female or male?* Toss a coin: Heads = FEMALE Tails = MALE
Stage 3	*Does the chick survive to leave the nest?* Take the next random number: 1,2,3,4,5,6,7 or 8 = YES 9 or 0 = NO # Go to stage 5
Stage 4	*Does the chick survive into the next two-year period?* Throw a six-sided die: 1,2,3,4 or 5 = YES 6 = NO
Stage 5	*Does the mother survive into the next two-year period?* Throw the die again: 1,2,3,4 or 5 = YES 6 = NO
Stage 6	*Does the father survive into the next two-year period?* Throw the die again: 1,2,3,4 or 5 = YES 6 = NO
Stage 7	Write the total number of birds alive now, 2 years later. **Note:** If either of the parents dies, and the chick is less than 10 months old, the chick will also die.

Each decision box in the flow chart requires a randomised results based on probabilities calculated from real data. The simulation follows this flow chart through the breeding cycle for one pair of birds. With simulation aids like a die, a coin, a 1-10 random-number generator, the path is chosen through the chart, based on the following relative frequencies.

- 70% of eggs will hatch
- 80% of chick survive to leave the nest
- 85% of birds survive over a two-year period.

6 STATISTICAL SIMULATIONS IN THE CLASSROOM

In these statistical simulations, probability and statistics appear as interwoven strands of the same theme, rather than as disjoint areas of mathematics. Real situations are studied, and real data is used.

Mathematics is applied in a real context, of interest to the pupil.

Statistical simulations provide the opportunity for cross-curricular projects. Situations can be modelled from numerous subject areas including ecology, biology, population studies, business applications, physics and mechanics.

Students are also motivated to write simple programs for graphical calculators or microcomputers.

REFERENCES

Hudson, B and Rouncefield, M (1991). *Handling Data Extension* Year 7/8. Century Maths Series. Cheltenham: Stanley Thornes.

Hudson, B and Rouncefield, M (1992). *Handling Data Extension* Year 9. Century Maths Series. Cheltenham: Stanley Thornes.

Rouncefield, M and Holmes, P (1989). *Practical Statistics*. London: Macmillan Education.

9

Interactive video as an aid to modelling

John Gillespie and Richard Phillips
Shell Centre for Mathematical Education, University of Nottingham, UK

SUMMARY

A government-funded contract in the UK has provided an opportunity for the development of interactive video materials by the Shell Centre and New Media Productions to support the learning of number at secondary level. This opportunity has enabled the developers to explore the use of this technology in the mathematics classroom in a variety of ways, including the generating of data and experience to do with modelling and simulation.

This paper gives the background of the project, some preliminary conclusions about the use of this technology, describes some of the modules produced, and discusses some resulting issues.

1 INTRODUCTION

The Shell Centre for Mathematical Education at Nottingham University and New Media Productions of London have collaborated over the last two years to produce an interactive video package to support mathematics learning at secondary-school level. This development project, *The World of Number*, was the outcome of a contract awarded by the (now) Department for Education to support the learning of number. The contract specified the use of Laservision discs and player, in conjunction with a PC and graphics board, a combination that is well tried in fields such as commercial and technical training but very little used in schools at present.

The World of Number package consists of 16 modules of very varied styles, produced on three double-sided laser discs with associated software and extensive accompanying print materials.

It is a feature of most of the work at the Shell Centre that trialling and working closely with schools has a central place, and this was true for the development of The World of Number. We worked with 12 schools, developing our ideas for later modules as we observed earlier modules in use in the classroom. Several of these modules – On the Move, Perspectives, Mechanisms and Powers of Ten among them – provide starting points for modelling and simulation activities.

We believe there are lessons that can be learned from the project concerning the development of materials for the secondary-mathematics classroom including the support of simulation and modelling. This paper highlights some of these issues, in the light of our experiences at the beginning of the 'multimedia revolution in the classroom'.

First of all, what did our platform of Laservision disc player and PC allow us to do? Each Laservision disc holds over 40 000 still pictures with sound. When run continuously, this gives about half an hour of high-quality video – running at 25 frames per second. The computer is used to control which of these pictures is on the screen. The screen can also display normal computer-generated text and images, on their own or together with the video images from the Laservision disc. Software-generated menus, with loops and branches, enable users to move between stills or picture sequences of their own choice. Users can click on 'buttons' on the screen to control the system, just as they do when using a normal computer.

Additional applications have been programmed as part of our project. Among these, Toolbox allow users to place markers on video images, draw lines and measure angles between them, and measure screen distances between markers.

When our project started over two years ago, platforms based on laserdisc players dominated the field. Whilst providing high-quality moving-video images, because these images and sounds are stored on laser discs, the images cannot be modified by the computer. Now with the development of more powerful computer technology, multimedia platforms are appearing where the pictures and sounds are stored digitally, so they can be accessed and modified like other digitally-stored data. Technical possibilies for developers of software are expanding rapidly, the quality of retrieved moving images is improving, while equipment prices are going down. Thus many more possibilities are now open to developers than was the case even a few years ago.

2 DESIGN OBJECTS

In most countries, including the UK, the contents of mathematics syllabuses are laid down, but the teaching approaches and the links made between sections of a syllabus are personal to the teacher. It is important that any multimedia package allows the teacher to retain this freedom, to enable the teacher to be supported rather than prescribed by the package.

In our project, the aim was not so much to aid the mastery of a particular skill (as one uses multimedia to train workers in a specific technique), nor to provide a self-contained experience, but rather to extend the normal active-learning environment of the effective classroom. The equipment becomes one of the normal classroom resources, used in many different ways and at many different levels.

Most classroom are very busy places. Teachers are typically doing several things at once, including deciding what comes next. So it is important that the teacher can move about the audiovisual material without delay, and be able to change route as classroom circumstances change. Elaborately planned pre-programmed presentations tend not to be flexible enough for most classrooms. After all, the teacher and the equipment should be enablers of learning, responding to the needs and questions of the students.

As a result of our experiences, we have reached some views on the character of multimedia support in the mathematics classroom. We believe that the multimedia units of material should be such that

- the teacher will find it easy and non-threatening to integrate the units into the normal work of the class
- students will learn to use the equipment easily and quickly
- units are 'open', and can be used to suit the user's needs rather than just those of the originator
- a single piece of equipment can 'service' a class of 25-30 students, working on several different problems
- a unit can act as a start or end point for work off-screen

and that multimedia units can help to

- support the learning of 'hard' topics
- extend the variety of approach
- bring the outside world into the classroom
- give new insights to topics by providing fresh viewpoints

3 DISCUSSION OF SOME APPLICATIONS

On the Move

This module includes about 20 short sequences of objects in motion, ranging from motorway traffic through Olympic athletes to the motion of plants brought to life through time-lapse photography. Just the ability to study the motion frame-by-frame enables data to be gathered and studied which would not be possible even with videotape. Opportunities for further analysis then suggest themselves, such as measuring the speed of traffic using regularly-spaced road markings as a method of measuring distance, and analysing the motion of sprinters, divers and throwers in athletics.

Mechanisms

Ideas of ratio and locus in linkages and lever systems are attractive but difficult topics, partly because – although everyday examples of mechanisms can be brought in to the classroom and observed – it can be hard to abstract data from them in order to do further work. We tried to tackle these problems in the unit on Mechanisms.

In Mechanical House users find themselves in a house with over 20 different everyday mechanisms coming into view during a short story sequence – a clothes drier, various domestic tools, umbrellas, an anglepoise lamp, a bicycle, and a box with trays hinged inside it. Users can click on a mechanism and then explore it more fully. It is up to the users to decide how to investigate, and to do their own calculations. The mathematics is not 'taken away ' from the students – rather, they are given access to a world which they can research as they decide. The teacher can make use of the suggestions in the printed material for particular tasks, if required.

Let us look at the box with trays inside it. The illustration shows one frame from the video sequences, together with some Toolbox marker points referred to below (Fig. 1). User can then bring up a still picture of it, positioned so that the plane of operation of the mechanism is parallel to that of the monitor screen. They can see the mechanism operate at normal speed, slow speed, or one frame at a time. It is then possible to investigate the mechanism in several ways, such as

a plotting the loci of different points as they move;
b comparing the distances moved by two points on the mechanism as the mechanism moves;
c measuring the angles turned through by different parts of the linkage holding the trays;
d seeing why the trays do not tip up as the box is closed.

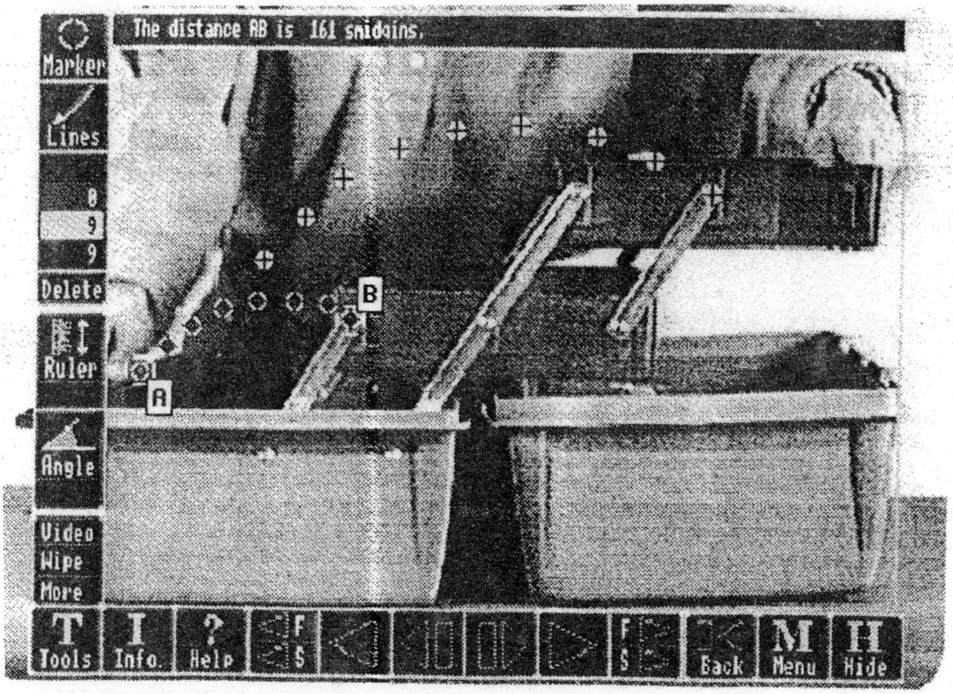

Figure 1. The box with trays. Marker points are already plotted.

For **a**, a marker dot is placed on a chosen point of the mechanism on the screen, using Toolbox. The film is moved on a few frames – that is, a few 25ths of a second – and another marker is placed on the new position of the same point. The process is repeated to produce a trace of the locus.

For **b**, two positions of a point are marked on the screen. Toolbox enables the distance between the positions to be measured (in screen units). The distance moved on the screen by another point can also be measured. Then the ratio of one distance to the other can be found.

For **c**, Toolbox enables angles to be measured between two lines.

For **d**, lines can be drawn along the trays and the links of the mechanism, and angles measured between them and compared.

The second illustration (Fig. 2) shows the matching support sheet.

Dimensions (diagram not to scale)

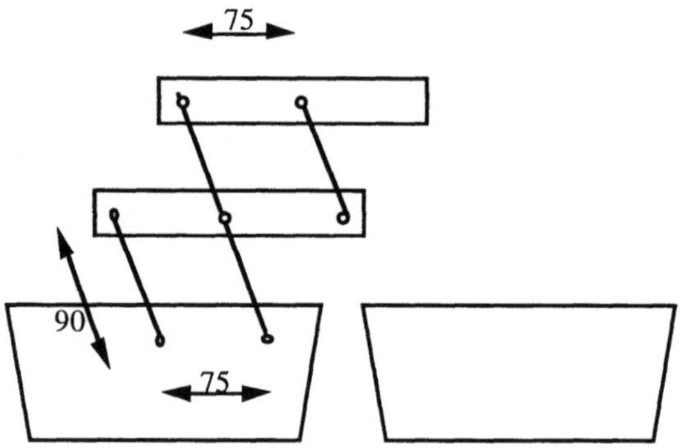

- See how the mechanism keeps the trays level.
  What features of the mecanism make this happen?

- Which other mechanisms is this like?

- How far does the trap move, compared with the bottom tray?

- Design a toolbox with 3 trays instead of 2.

Figure 2

In similar ways, the loci of any moving points chosen by the user can be plotted, and the relative distances moved by several points can be measured and compared, providing data for ratio and mechanical-advantage investigations. Compared to analysing the actual objects in motion, Mechanical House allows movement to be studied very slowly – in 25ths of a second – and the same sequence to be accessed as many times as desired.

Thus the user can work in a practical and investigative way with aspects of mathematics such as

- posing problems
- putting forward hypotheses and testing them
- working with ratio
- working with decimals and fractions
- working with loci
- working with angles and parallel lines
- working with linkages and other mechanisms involving gears
- using ideas of leverage and mechanical advantage.

Much more work in the field of mechanisms in the mathematics classroom has been carried out by members of the Mechanics in Action project at the University of Manchester.

Powers of Ten

Another difficult area of mathematics is the understanding and representation of very large or small quantities. It is difficult to bring Standard Index Notation alive in the average classroom on a Friday afternoon!

Many will be familiar with the film Powers of Ten and its associated book. We were delighted to be able to include this graphic film in our package. It provides a vivid but essentially a spectator's view of our world and its setting in the universe. The viewer sees a picnic by the lakeside in Chicago, then the viewpoint moves upwards from the picnic so that, after every ten seconds, the distance from the picnic is multiplied by ten. At the end of the outward journey the viewer appears to be at the edge of the universe.

Our software allows users to measure any chosen distance in space that appears on the screen (through a plane passing through the picnic, to be precise) using Standard Form. A development allows users to go on journeys of their own, by multiplying distances by numbers in standard form. So the equipment enables users to set problems of their own and work on them.

Perspectives

Using a selection of the Olympic clips in On the Move, one section of the Perspectives module presents users with 'stacks' of graph scales and graph lines, from which the users make a selection to match a particular example of motion. The correct selections result in the complete graph being plotted on the screen as an overlay, in step with the motion. In a vivid way users see how their chosen graph relates to the motion, as the actual motion takes place.

4 MULTIPLE LEVELS

We have found that the same section of a module can support mathematics at several different levels, or even different mathematics topics. For example, with several of the mechanisms students could look at a still and be asked to suggest how different parts might move; they could then watch the mechanism in motion and go on to suggest how it might be modified, perhaps involving the making of models of their ideas.

At a higher level, students might observe the changes in mechanical advantage, or movements of centres of gravity, which occur as the geometry of the mechanism alters, investigate reasons why a particular design feature had been included, and explore the effects on properties of the mechanism of modifications in design.

An expensive investment for a typical school can thus be used to maximum effect. In the first three of the applications above no questions are set, so the teacher and students are free to pose and investigate problems at a wide variety of levels.

5 THE ROLE OF PRINT SUPPORT MATERIALS

We have developed extensive print-support materials to use alongside the system. Many of these consist of teaching ideas and worksheets. It is up to the teacher to decide whether to use a particular set of worksheets – so teachers can select those sheets which particularly suit their class, or include their own. The print materials also include a detailed index and contents listings, together with operating instructions.

Had the audiovisual system had sets of questions built in throughout, there is the danger that control of learning passes from teachers and their students to the designer of the software.

Some of the modules set up particular problem situations. Even here, though, the problems have been chosen so that they can be extended by the students, with print materials (from the pack or produced by the teacher) on hand if required to support this process.

6 THE ROLE OF TRIAL SCHOOL TEACHERS AND STUDENTS IN THE DEVELOPMENT PROCESS

We were lucky enough to build up close working relationships with our 12 trial schools. In fact some of the schools provided actors for filming. Throughout we tried to keep any live presenters and voice-overs as close to the potential student users as possible. There is no screen 'expert' for most of the time. This was a hunch which appears to have paid off. Less than perfect acting is more than compensated for by the ease with which our student users identify with the screen characters.

7 EVALUATION PHASE AND CONCLUSIONS

Our preliminary conclusions from our own trial and development experiences are that multimedia can be assimilated into the classroom quickly, that students in particular quickly become adept at using the apparatus, that it can perform a significant function in supporting mathematics learning and complementing other resources, and that it opens up many new perspectives in mathematics, including aspects of modelling and simulation, quite apart from any specific skill-training function.

Currently about 30 schools thoughout England and Wales are embarking on a year-long Department for Education evaluation of multimedia. Our package forms part of this wider trial.

We look forward to learning more about the possibilities and opportunities for multimedia as a genuine support for learning, from this evaluation and from further work of our own.

REFERENCES

Gillespie, J A (1992). *IV and Mathematics.* Time Europe 1992 Conference Proceedings. Eindhoven: EPCO.

Hoffos, S *et al* (1992). *CD-I Designer's Guide.* Maidenhead: McGrawHill.

Kitchen, A *et al* (1992). *Made to Measure! Linkages.* Century Maths. Cheltenham: Stanley Thornes.

Morrison, P (1982). *Powers of Ten.* New York: Scientific American Library.

Steeg, T *et al* (in press). Using the Geometer's Sketchpad to work on linkages: some preliminary experiences in mathematics and technology classes. In *Micromaths.*

Williams, J S (1991). Modelling in mathematics: a cross curricular and problem solving approach to learning mathematics. In Niss, M, Blum, W and Huntley, I (ed), *Teaching of Mathematical Modelling and Applications.* Chichester: Ellis Horwood, 271-279.

10

Learning activities involving mathematics in real-life situations

Paulo Abrantes
University of Lisbon, Portugal

SUMMARY

After three years of attending an innovative curriculum for mathematics – which emphasised application-oriented activities in the classroom, especially situations organised as project work – 20 students of grade 9 participated in a competition which required the use of mathematics in a realistic situation. An analysis of their performance, when compared with that of students from other classes, suggests that some cognitive and noncognitive factors needed in such situations can be related to their previous experience and seen as *learned behaviour*. This paper tries to point out the most promising results in that respect, as well as to identify major limitations still observed.

1 INTRODUCTION

From 1988 to 1992, the MAT789 Project (1) built an experimental curriculum for mathematics in the lower secondary level (grades 7-9, pupils aged 12-15), working with four classes of two schools in the Lisbon area. The Project received institutional support from the University. An authorisation from the Ministry of Education was necessary, because Portugal has a centralised school structure with rigid national curricula and programmes. The Project team included two researchers and the three mathematics teachers of the experimental classes. These five members met every week to work on all aspects of the curriculum.

The Project maintained that school should, above all, contribute to develop understanding of the rôle mathematics plays in each student's life and in society, and to generate positive attitudes towards mathematics. The curriculum intended to emphasise processes rather than factual knowledge, and ideas rather than techniques, and to promote problem-solving experiences involving relations between mathematics and various fields of reality as well as internal to mathematics. Students should have the opportunity and time to develop meaningful activities, and to reflect on them. A variety of working processes was considered, including small-group work (the most frequent classroom setting), elaboration of written reports and comments, discussions, personal study and research, practical work and project work. The extensive use of calculators, computers and concrete materials was also a characteristic of the curriculum, as was a considerable change in the assessment methods which considered the large variety of students' productions together with two-stage tests (Leal and Abrantes, 1992).

Most students' activities can be grouped into three main types: (a) thematic sequences – questions and problems about a topic or related topics; (b) situations or open-ended problems to explore and to comment on; (c) project work – students developed two or three projects each year. Relations between mathematics and reality played a crucial rôle in the curriculum, being the starting point for many thematic sequences and the focus of most open-ended problems and all the projects. For some more details and examples see Abrantes (1991).

All these aspects strongly contrast with the national mathematics curriculum in Portugal which promotes an individualistic, formalised and content-centred teaching of mathematics. One of the major differences relates to application-oriented activities. In the MAT789 curriculum these have been organised according to the 'integrated approach' (Blum and Niss, 1991) and independent work done by the students was a necessary component, while in the national programmes applications are limited to illustration of concepts, and project work does not exist at all.

2 THE *MATHEMATICS AND REALITY* CONTEST

In May 1992, the MAT789 Project team organised a competition entitled *Mathematics and Reality*. Seventy nine graders coming from four different schools participated. They included all 20 students belonging to one of the MAT789 classes, and 50 students with good or very good scores in mathematics coming from other classes. The contest was organised according to rules which are not usual in mathematical competitions:

- the students were organised in small groups of 3 to 5;
- the work consisted of producing an answer to a single problem situation;
- the situation referred to a real-life problem that could be interpreted and explored with the support of mathematical processes, and took the form of an open-ended problem – for which there is not an approach or a solution previously considered as correct or better than any other;
- students had a whole day and a free atmosphere in which to work, and they could consult and use all the available material, including books, calculators and computers.

Study the following problem and propose your solutions. In a short time and under the conditions of the contest you do not have possibility to collect all the needed data. However, you must describe the processes you would use to get those data. As your proposal must be as concrete as possible, you may invent the data you will need to explain how your model works.

The avenues D. José I and Dr. José Pontes allow two-way traffic flows. Near the crossroads, the first is wider and has three lanes, because the traffic is very intense. **To regulate the traffic at this crossroads, traffic lights could be installed. But there are various problems to solve and aspects to be considered...**

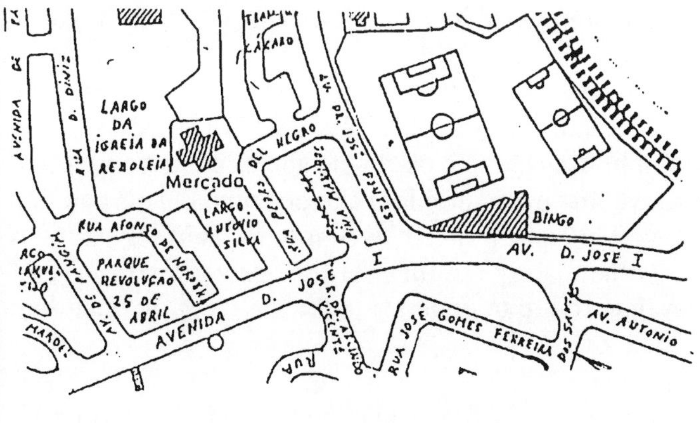

- The places where the traffic lights should be installed.

- The best way to organise the system: in a given moment, some lights are *open* while others are *closed*; in the following phase, some *close*, some *open* and others remain as they are...

- The period of *openness* is not necessarily the same for all lights or for all periods of the day... In a given period how could (should) the system work?

- The decision to install a system of traffic lights costs money. The opinion of the interested people should not be neglected.

And there are probably other relevant aspects...

Figure 1. Reproduction, in a smaller size and in an adapted English version, of the worksheet that the students received for the contest.

The contest had been thought of initially as a way to assess how the students from the MAT789 class would respond by their own to a realistic, unknown problem. However, the organisers decided to invite other classes, having in mind the additional objective of generating some attention in the kind of problem usually undervalued in the mathematics classrooms.

The proposed problem focussed on a situation related to the region where the students lived. They were invited to produce a proposal – to describe and to justify it – to create a system of traffic lights at a crossroad (see Fig. 1).

3 KNOWING MATHEMATICS – AND KNOWING HOW TO USE IT

Every experienced teacher knows students who find it very difficult to solve new problems, sometimes very simple problems, even when they are familiar with all the necessary procedures.

Several researchers have conducted studies on the processes used by good problem solvers. Schoenfeld (1987) analyses the ways that a geometrical problem was approached by students who knew the mathematics needed to solve it but were not able to do it, and a mathematician who had not worked in geometry for a number of years but actually succeeded in solving the problem. Schoenfeld states that the difference lay in how they "made use of what they did know". Contrary to what the students did, the mathematician spent the majority of his time *thinking* rather than *doing,* and he used efficient self-monitoring and self-regulation.

These metacognitive skills are not the only characteristics of good problem solvers. The same author emphasises the rôle of beliefs about mathematics – for example, students who believe that every problem can be solved in 10 minutes or less will tend to give up working on a problem even if they could solve it with more effort, or those who value mainly the *form* of the answer tend to spend more time with that aspect than trying to understand the *meaning* of what they are writing.

Even when the students have learnt the content knowledge *and* the appropriate cognitive processes it is not automatic that they will use them. Resnick (1987) says that learning to recognise – and even search for opportunities to apply – one's capacities is essential, and that schools must cultivate not only thinking skills but also the disposition to use them. Moreover, she admits that "some correlation exists between the kinds of motivations that keep people working and several qualitative features of their work".

It seems reasonable to suppose that these factors have to do with a large variety of problems, including those which involve application of mathematics. However, it should also be admitted that using mathematics in realistic situations will require specific skills coming from the nature of those situations, and not generally developed by *purely mathematical* problems.

For example, many realistic problems are not well-defined and ready to be answered. This fact may cause a lot of difficulties – students not familiar with this kind of problems tend to lose confidence because they do not know what they are supposed to do. It makes sense to discuss the ability to formulate problems in a way that will make them tractable by known mathematical processes.

Another characteristic of realistic problems is that "they often involve both too much and not enough information", to use the terms of Lesh (1990). Contrary to what happens in most problems in the mathematics classroom, it is necessary to ignore data which are irrelevant to the way we intend to approach the situation. At the same time, it may be necessary to collect new data, by consulting documents or people, or by using experimental methods.

Normally, we do not consider the reality itself, but a simplified version. Then, we translate the elements of that real model into mathematical terms, building a mathematical model to which we can apply known methods. After this, we have to confront the results with the real situation, and decide whether or not it is necessary to make corrections in the model, for example introducing factors which were previously ignored. Many authors distinguish between this *modelling cycle* and *mathematisation* – for example, Niss (1987) uses the latter to name the process of translating a situation into mathematical terms, that is, one of the steps of the cycle just mentioned.

Another source of difficulties in realistic problems originates in the necessity of considering extra-mathematical factors which play a decisive rôle in the real life situation. In the classroom, students often think that they only need the help of their teacher, or some other expert, in mathematics. In real life, however, problems are much more interdisciplinary. Moreover, as Lesh (1990) points out, if the students are familiar only with artificial problems in which 'math answers' are not the same as 'real answers' – and the differences are never discussed – they tend to suspend their real-world knowledge in the mathematics classroom.

Lesh quotes a textbook, showing how students are *encouraged* to do that: "Pat could run 8 miles in an hour; how far could she run in 10 hours?"

Real-life problems are normally open-ended – there is not a single way to solve them and, even when an approach is chosen, many decisions are still to be taken. In addition, the quality of the answer may be related to the explanations on the processes rather than to the results. Pollak (1969) says that these problems are often of the form "here is a situation, describe it (in a simplified or more meaningful form)!" rather than being of the form "here is a question and some data, answer it!".

Students' attitudes and beliefs also play an important rôle in their ability to solve realistic problems. These problems may require time and persistence, even if very specific knowledge of content topics is not needed. This will cause difficulties to those who only practice 'one-step problems', or those who will declare that a problem is impossible if it cannot be solved in a short time.

Are our students prepared to deal with all these difficulties? Does school mathematics allow students to develop the skills and attitudes needed to overcome them?

4 COMING BACK TO THE CONTEST

In the *Mathematics and Reality* contest, many students showed a remarkable creativity, and a disposition to carry out independent work. However, the lack of experience in this kind of activities was notorious. Several groups needed a lot of time to organise and begin to work seriously. Some of them did not discuss the solution in a collective way, starting very rapidly with practical tasks and ending the report in a short time. One group of good students wrote that they did not understand "what mathematics should be used", presenting a simplistic and incomplete work. Some groups were concerned only with the form of the final report.

Although the problem situation did not seem to be too complex, many groups presented traffic-light systems that would be impossible to implement in practice, either because they were totally irrealistic or because they simply were 'wrong' – in the sense that they would certainly provoke accidents. In addition, the quality of the explanations was also poor in several reports.

The five groups belonging to the MAT789 experimental class showed common qualities (also present in some of the other 12 groups but only in some of them):

- the capacity of organising the group and cooperating,
- persistence – contrary to what half of the other groups did, they all finished their reports near the time limit,
- strong intrinsic motivation, even from those who did not expect to win one of the prizes.

These results are already interesting. However, a closer look into the assessment criteria and the characteristics of the best work will give further indications.

The jury was composed of five teachers from different schools, and the students' reports were anonymous before the decisions were taken. The jury had to take into account criteria such as the richness and practicability of the proposed solution, the relevance and correctness of the mathematical aspects involved, the quality of the argument, and the organisation and originality of the work. Contrary to what occurs with objective tests, assessment of this kind of report tends to be:

- more absolute than relative – each work has to be confronted with general criteria but also with criteria specifically directed to the way how the group approached the situation,
- more qualitative than quantitative,
- more holistic than the result of the sum of the separate skills.

The jury's main task was to decide who would receive the first and second prizes. However, in these situations unpredictable facts occur. Some reports could not be considered the best, but they were very good in a particular aspect. The jury decided to attribute special mentions, like in the Oscars of the cinema. For example, one of them was for *originality*, won by a group who presented a nice three-dimensional model and a poster as final report. Unfortunately, the proposed solution was not adequate in some phases of the traffic lights system, and the explanation was not very clear.

5 THE BEST WORKS

As a first step towards the final decision, the jury selected four reports as being the best. All of them proposed three phases for each cycle of the system, and correctly stated realistic periods of time for each traffic light to be green.

Two of these reports got special mentions, but they did not win the first or second prize because they failed some important criteria.

- One of them won a special mention for *argument,* because it contained the best schematic explanation about the proposed system. However, this group did not respect the real conditions, artificially widening the roads to make their solution possible. They considered three lanes in one of the avenues and four in the other. Moreover, they did not pay enough attention to the pedestrians, only saying that "they must cross when cars are stopped".

- The other report won a special mention for *planning,* because it presented the most complete sequence of steps to approach the problem, starting from a study about the number of cars passing in the several directions as a basis to further decisions. However, the solution was too simplistic, without any reference to problems such as the various ways for the pedestrians to cross the roads.

The two remaining works won the first and second prizes (Fig. 2 refers to the proposal of the winning team). They both have been produced by students coming from the participating MAT789 experimental class. In common, these two reports consider the actual conditions of the situation, namely the facts that the traffic is more intense in the wider avenue, that it is possible to have three lanes in this avenue, and that 'direction Lisboa' requires priority. Furthermore, these two reports were the only ones that consistently took into account the needs of the pedestrians – one of them proposed an aerial passage above the wider avenue, and the other included two additional lights at strategic points.

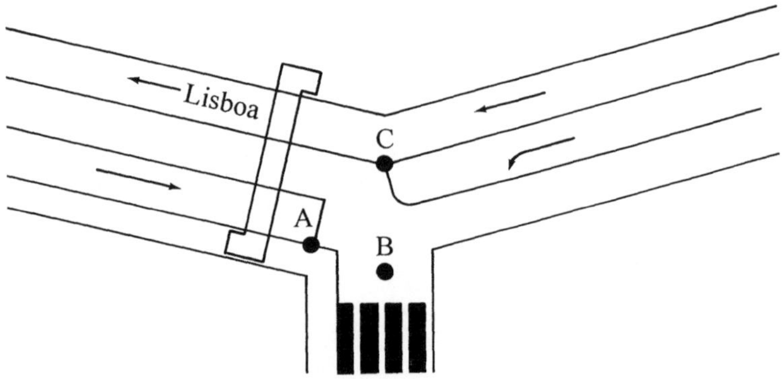

Figure 2. Drawing presented by the winning team. Lights A, B and C are open for 45, 15 and 30 seconds, respectively. Cars following 'direction Lisboa' do not need to stop. The system is practicable and realistic.

Observation of the way that students worked, analysis of the reports, and interviews with some of the students revealed that these two groups:

- dedicated time to discuss and plan, before starting the practical tasks ("first, we tried to go into the structure of the problem, to think of every road, every direction for cars, where we want to cross, and where accidents could happen – what were the problems and what was necessary to be done");
- adopted a realistic perspective and even tried to use their experience as pedestrians ("since we cross that avenue every day we know there are some problems that need to be solved");
- showed a willingness to take risks, incorporating factors that could become a potential source of errors but that would improve their work;
- used a variety of materials, including the computer, and all the available time (the winner was even the last group to finish);
- paid considerable attention to communication of the proposal, both to the text and the adequate drawings.

It is remarkable that the best works were produced by average or even weak students, when their scores in school disciplines are compared with the scores of most of the others.

Case studies aiming to investigate the evolution of some of these students in grades 7, 8 and 9 (one of them is reported in Abrantes, 1992) suggest that their cognitive skills and attitudes about realistic problem solving can be related to their mathematics curriculum. For example, a girl who played an essential rôle in the production of one of the best reports declared "For a student of a normal class, it is difficult. In our class, he would not be able to do many things because he only knows the mathematics itself. He does not do mathematics in practice as we do". In grade 7, this girl had a lot of difficulties with the vagueness of new problems, and lack of confidence in the use of mathematics when she had to work independently. However, three years later she spontaneously points out project work as her favourite activity, and something that should be done in every class, at least once a year. At the end of grade 9, she says that her "strongest point" in mathematics is "a kind of approach to a situation, reasoning, in general not in a specific topic" and she adds "Mathematics is reasoning more than any other thing, how to present, how to say... the kind of things that can be applied to other subjects".

It appears that cognitive and noncognitive factors related to the use of mathematics in realistic situations can be seen as *learned behaviour*. This seems to have progressed with continuous student experience on activities developed in the experimental mathematics curriculum, particularly in those areas which were organised as projects on realistic problems.

However, some weaknesses could also be observed in all the participants in relation to a few aspects. One of them is the intellectual quality of the discussions that took place among the members of a group. Very often, the arguments were too short, and focussed only on practical points; each student was mainly interested in imposing their own solution rather than carefully listening to different points of view. Although students who were familiar with small-group work showed improved skills related to cooperation, an effective progress involving both cognitive and noncognitive aspects seems to be a difficult and long-term process.

Another weakness is related to the explicit importance given to the discussion of the mathematical model which is being developed. In this respect, large differences could be observed among the groups, as was pointed out earlier, but even the best groups seemed to be too much attracted by the form of the report. Maybe the competition format of the activity favoured this attitude. However, at this age level and at this low level of mathematical knowledge, it seems difficult – but necessary – to help students to develop the practice of taking the discussion of the model as a major aspect of their work.

6 FINAL REMARKS

Just as we can expect that students with an intensive training in solving equations would probably win a contest consisting of rapidly solving a lot of equations, it is not surprising that students more experienced in developing projects which involve the use of mathematics in real-life situations are more familiar with important components of the work on such situations. However, the way to develop corresponding skills and attitudes is different, as is the way to assess them. To determine if a student is able to solve an equation, we can change the coefficients in one that he has already solved; to assess the skills to solve a new problem we would not change the traffic lights of the previous project to a policeman...

The school can – and should – consider the development of various skills and attitudes needed to use mathematics appropriately in real-life situations. That requires specific teaching, and attention to particular features and difficulties of those situations, and adequate learning activities. Although we should not deny the value of artificial or structured problems, students should be confronted – at least sometimes, as Niss (1989) says – with authentic problems, participating in the entire modelling process, and working in an active and independent way. In this sense, project work as a learning environment to interprete or solve realistic problems may play a very important rôle in the mathematics curriculum.

It is worth repeating that the main limitations of the situation described in this paper came from its contest format. Under different circumstances, the problem could lead to an extended and more realistic project, and some of the difficulties with assessment would disappear because it would not be necessary to decide who was to receive the prizes. Since the ability to use mathematics in real-life situations is a relevant aspect of this science, as well as the major reason for its importance in school, it seems reasonable to point out the mathematics classroom as a place to pay attention to the development of such ability.

NOTE

(1) The MAT789 Project team includes the author of this paper and also Eduardo Veloso, Leonor Cunha Leal, Margarida Oliveira and Paula Teixeira. The Project is funded by the *Calouste Gulbenkian* Foundation.

REFERENCES

Abrantes, P (1991). The rôle of applications in a curriculum project for school mathematics. In Niss, M *et al* (ed), *Teaching of Mathematical Modelling and Applications*. Chichester: Ellis Horwood, 128-136.

Abrantes, P (1992). Project work in school mathematics: an experience in Portugal. In Leino, J, *Mathematics Teaching Through Project Work.* Hämeenlinna: University of Tampere, 42-48.

Blum, W and Niss, M (1991). Applied mathematical problem solving, modelling, applications, and links to other subjects – state, trends and issues in mathematics instruction. In *Educational Studies in Mathematics,* **22**, 37-68.

Leal, L and Abrantes, P (1992). Assessment in an innovative curriculum project for mathematics in grades 7-9 in Portugal. In Niss, M (ed), *Cases of Assessment in Mathematics Education: An ICMI Study.* Dordrecht: Kluwer Academic Publishers, 173-182.

Lesh, R (1990). Computer-based assessment of higher order understanding and processes in elementary mathematics. In Kulm, G (ed), *Assessing Higher Order Thinking in Mathematics.* Washington: American Association for the Advancement of Science, 81-110.

Niss, M (1989). Aims and scope of applications and modelling in mathematics curricula. In Blum, W *et al* (ed), *Applications and Modelling in Learning and Teaching Mathematics*. Chichester: Ellis Horwood, 22-31.

Pollak, H (1969). On some of the problems of teaching applications of mathematics. In *Educational Studies in Mathematics*, **1**, 24-30.

Resnick, L (1987). *Education and Learning to Think*. Washington: National Academy Press.

Schoenfeld, A (1987). What's all the fuss about metacognition? In Schoenfeld, A (ed), *Cognitive Science and Mathematics Education*. Hillsdale: Lawrence Erlbaum.

11

Mathematics of track and field events – Composite Real Mathematics approach

Akira Yanagimoto
Tennoji Junior High School, affiliated with Osaka Kyoiku University, Japan
Yuichi Mori
Kurashiki City College, Okayama, Japan
Tetsuo Matsumiya
Osaka Kyoiku University, Osaka, Japan

SUMMARY

This paper is an extension of the method called Composite Real Mathematics (CRM) whose purpose is to give solutions to some problems in mathematics education which are encountered particularly in the lower-secondary schools in Japan. This method is utilised in the material we have developed called Mathematics of Track and Field Events, where the students' pre- and post-test scores were recorded. From this study, we were able to show the validity of the material used, the application and usefulness of computers, and the differences in students' overall performance.

1 INTRODUCTION

We feel that there are several aspects of mathematics education in the present educational system of Japan that require wide discussion in order to improve the learning process. The mathematics curriculum used in lower-secondary school is based on the syllabus furnished by the Ministry of Education of Japan, but this often implies that the purpose of the traditional learning process is to acquire knowledge and skills consecutively inside the classroom. Often the students learn the different mathematical ideas and theories without the rationalisation of applying them. This implies that the content of the mathematical courses offered in the lower-secondary schools becomes increasingly abstract and theoretical in nature. This situation leads to a strong phobia, among students, against entering upper-secondary mathematics classes. The mathematics problems in the entrance examinations, though original, may not be related to the real world. Some of the problems involve commonly-stressed computations, and students have little opportunity for real understanding and development of their reasoning abilities. Nowadays, it has become common for students to enroll into cram schools called Juku, where students are taught the techniques to solve this kind of problems in entrance examination. As a consequence, most students find the problem-solving item mere routine, and not related to the real world. These are some reasons why most students develop a negative attitude to mathematics, as they move from grade to grade. Thus modelling and real-world examples play only a minor rôle in the classroom.

Reflecting upon the situation mentioned above, the new syllabus in 1993 has included various applications of mathematical concepts, including problem solving, in the curriculum of the lower-secondary schools. At the beginning of the 1980s we developed the Composite Real Mathematics approach, that features the following aspects (Matsumiya, Yanagimoto and Mori, 1989; Matsumiya, 1991).

- To formulate a suitable teaching strategy for the existing mathematics curriculum, for solving realistic problems.
- To choose problems which satisfy the conditions
  - to be realistic,
  - to be handled by students' actual activities (including using computers),
  - to be solved with several different ideas of mathematics,
  - to be treated continuously in a limited time (less than 10 hours).
- To place emphasis on the process of mathematising, modelling and practical application, rather than mere computation.
- Sometimes to separate the students into small groups.
- To utilise this approach two or three times a year.

We were able to relate the CRM approach, viewed as an international perspective, with the international trends of applications and modelling in mathematics teaching, as discussed in the papers of Niss (1989), Blum and Niss

(1989) and Blum (1991). We have already developed some examples of the CRM approach – see Mathematics of a Lake (Matsumiya, Yanagimoto and Mori, 1989), Mathematics of a High Building (Yanagimoto, 1991), and Mathematics of a Railway (Matsumiya, Yanagimoto and Mori, 1992). Moreover, it is worth mentioning that this approach is similar to the mixing and integrated mathematics approach of Blum and Niss (1989). However, the full integration between mathematical and extra-mathematical activities, within an interdisciplinary framework, has not been incorporated into our approach. From the viewpoint of the two trends – pragmatic and scientific-humanistic – in the international debate on applications and modelling (see Kaiser-Messmer, 1991), our approach includes both trends but is closer to the latter.

2 TEACHING APPROACH

Materials and resources

We made use of the mathematics of track and field events as our major source of material satisfying the different conditions discussed above. Since these types of sport have recently been gaining in popularity among students, this gave us a strong motivation to analyse them scientifically. Table 1 shows the actual lesson plan that was administered to the third-year students in the lower-secondary school with 14 to 15 year old students.

Unit	Content	Alloted time	Mathematical tools	Activities or visual aids
I	designing of an athletics track	3 periods	length of circular arc, Pythagorean theorem	computer simulation (teacher-developed), calculator
II	record of running broad jump	4 periods	correlation, quadratic equation	correlation diagram, calculator, application software (spreadsheet)
III	location of the markpoint in passing a baton	2 periods	linear function, quadratic function quadratic equation	drawing graphs, calculator

Table 1. Lesson plan of Mathematics of Track and Field Events

The lessons were held from early 1991 over 10 periods, each period lasting for 50 minutes, for 159 third year students of Tennoji Junior High School affiliated with Osaka Kyoiku University.

Assessment of students' overall performance

Here, we administered two types of test to measure and evaluate the students' overall performance in their activity.

- Pre-test (about 15 minutes) covers
  - students' recognition of mathematical concepts, theories, etc in lower-secondary school,
  - students' preferences in mathematics.

- Post-test (about 30 minutes) includes
  - the same overall areas as the pre-test,
  - the students' comprehension of mathematical ideas applied in the Mathematics of Track and Field Events,
  - the students' interest in the contents of the material,
  - the students' impressions and personal views.

3 RESULTS OF THE TEACHING

Features of Unit I

The main features that were considered in Unit I are as follows.

- The main problem here was to design an athletics race track that will fit the space available – the school ground does not have enough room for a 400 m race track. Every student is aware of this problem, since each participates in the track and field races that are held every year.
- As reference material, every student is furnished with a plan of the school as well as the school ground.
- Personal computers and calculators are used wherever appropriate.
- The mathematical tools that are applied to solve the problem include the basic properties of circles, the Pythagorean theorem, and related mathematical ideas.
- The learning process of the students is classified into four main steps – knowing and understanding the problem, discussion of the methods of solution of the problem, finding the solution of the problem, and application and verification of the result.
- The students are divided into small groups, each composed of five to six members, and every group is provided with a personal computer.

Methods and strategies for solving the problem

Knowing and understanding the problem

The students, with the teacher in charge, discussed the strategies of attacking the problem. In the discussion they found and confirmed the following. In designing a race track, they need to consider the following data – the total distance of the circuit, the length of the straight part of the race track, the number of lanes, and the width of each lane. The construction of an athletics track is often done so that only the first half of the race circuit is provided with separate lanes, and the remaining part is a single track. In this way, different race courses produce varying shapes of track, and students need to calculate the best starting point for every race.

Discussion of methods of solving the problem

The mathematical ideas applied to constructing the athletics track include the following – radius of circular part $r = \frac{l - 2s}{2\pi}$, maximum horizontal length $H = 2r + s + 2nw$, and maximum vertical length $V = 2r + 2nw$, where l is the circuit length, s the length of the straight parts of the circuit, n the total number of lanes, and w the width of each lane (see Fig. 1). In order to construct an appropriate race track, students calculated the difference between the distances run in different lanes, expressed as $\pi(r + d) - \pi r = \pi d$, where d is the distance between the first and any other lane, and the difference caused by the single track, given by $B'C - BC = \sqrt{BC^2 + d^2} - BC$.

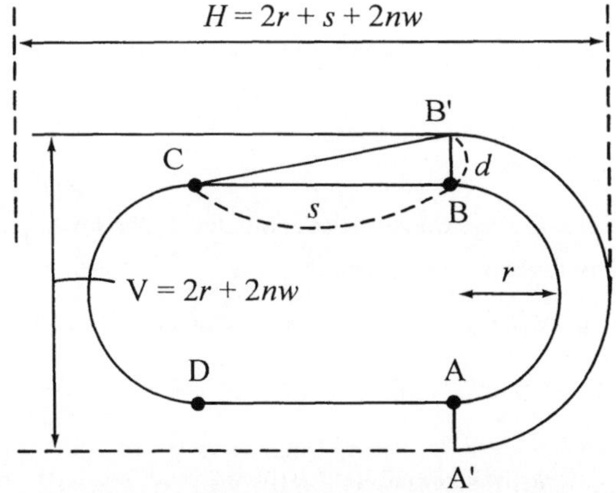

Figure 1. Differences between race tracks

Photo 1. Modelling a track on computer

Finding the solution of the problem
Following the above strategy, the students ran a computer simulation utilising computer software we have developed. Firstly they drew the coordinates of the outline of the school ground, based on the plan of the school. Then, after feeding in the different values of l, s, n, and w, the resulting shape of the race track was displayed on the screen. These steps were repeated several times, using trial-and-error to find the desired race track (as illustrated in Fig. 2). Some of the criteria of the optimal race track used by the students included the following. The preferred distance of a circuit should be 200 m, 250 m, or 300 m (as big as possible in 50 m increments); the distance of the straight part of the circuit must be proportional to the curve; a part of the ground should be intended for audience usage. Finally they calculated the differences between the race courses, and marked each starting point, as illustrated in Fig. 3.

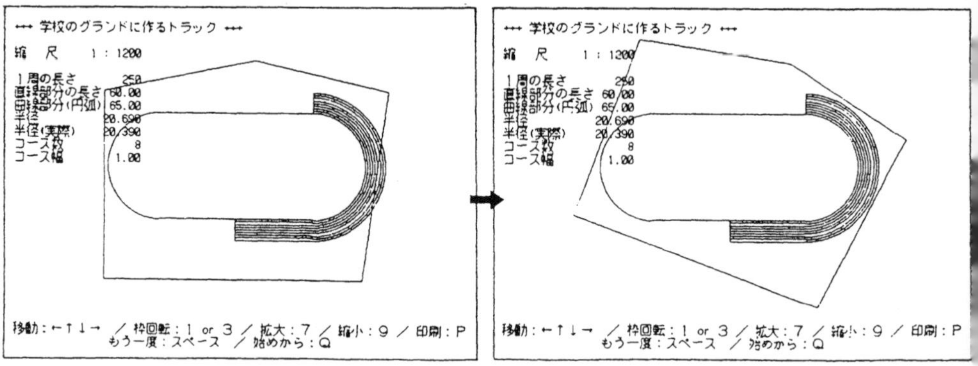

Figure 2. Trial-and-error method of designing a race track on the computer. In the lower part of the display, the following key functions are indicated: move (←↑↓→), rotate (1 or 3), magnify (7), reduce (9), hardcopy (P), again (space), and quit (Q).

Verification of the result
To check the validity of the result, a comparison was made with the results of the other groups, and the optimal one was chosen.

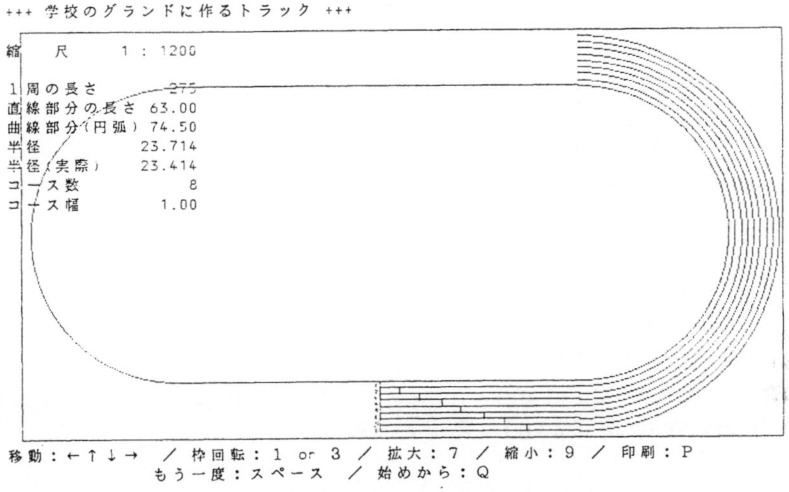

Figure 3. A sample of a hardcopy, illustrating the difference between race tracks. Input values: scale of display output=1:1200; l=275; s=63; n=8 and w=1. The rectangular frame shows the area of the school ground.

Unit II

A similar method of teaching was also applied in Unit II, to study the long jump. First, the students were divided into groups of four, and a computer was provided to each group. Then, the teacher prepared a record of every students' sports ability. Using these data, students were asked to select those sportif abilitiets which were strongly related to their ability in the long jump. With the use of a spreadsheet, by drawing correlation diagrams, they found that the 50 m run and the high jump were strongly related to their ability to do the long jump.

To calculate the running distance for the long jump, the students, with the help of the teachers, derived the formula $L = \frac{100}{s}\sqrt{\frac{2h}{4.9}}$, where s denotes the time of the 50 m run and h is the height of the high jump. Then the students were asked to make L close to their own record by adjusting the values of s and h. The result obtained was compared with data in the previous correlation diagrams. Finally, discussion followed on how to improve their ability for the long jump.

Unit III

In this unit, the students were asked to find the best track marking for passing a baton in a race such as a 400 m relay. When a preceding runner has reached the *Markpoint,* a succeeding runner should start running immediately, as illustrated in Fig. 4. This type of problem creates a realistic and interesting impression among students, since they often experience such activity during their annual sports competition.

To solve this problem, we selected two runners at random. The students approximated the distance of the preceding runner by the linear function $y = ax + b$, and that of the succeeding runner by a quadratic function $y = ax^2$, where x denotes the time and y represents the distance. Then, the graphs of the given functions were drawn by the students.

Next, the selection of the best location, or markpoint, was made using graphs and formulas. For these data, the quadratic equation obtained was given by $2.3x^2 = 7.7x - k$, as is illustrated in Fig. 5. Assuming that the discriminant of the quadratic equation is zero, we can solve this system for the best value of k.

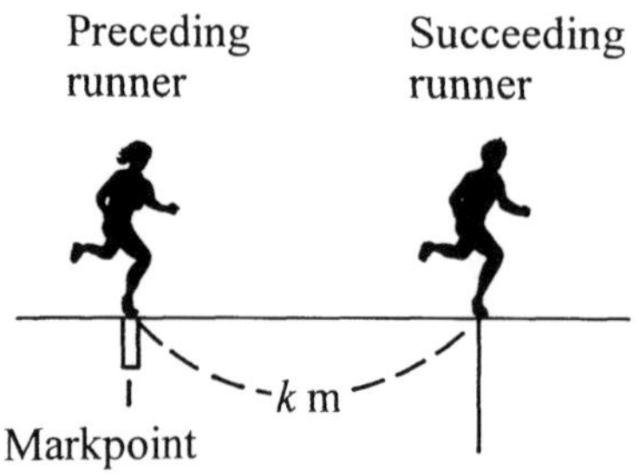

Figure 4. Best markpoint to pass a baton

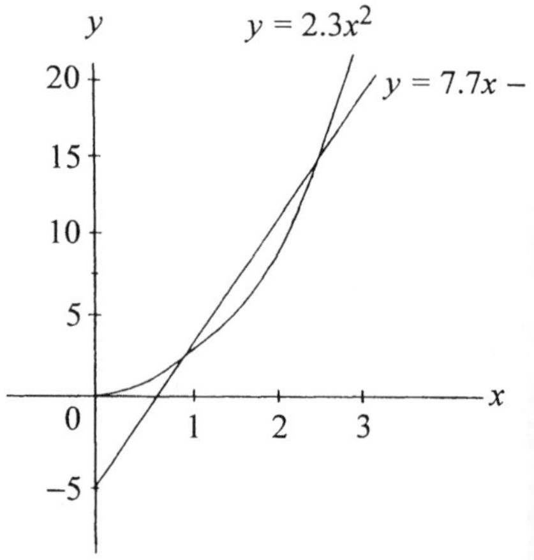

Figure 5. The case of a 5 m markpoint

4 EVALUATION OF STUDENTS' PERFORMANCE

Mathematics in the lower-secondary school

The assessment of the students' feelings about mathematics was done by administering a test before and after the lesson. The questions, which were the same for both tests, were expressed in the following form.

a Do you think that mathematics is related to activities in your daily life?
b Do you think it is necessary to study mathematics through the use of actual activities such as experimentation, computer manipulation, and suchlike?
c Do you usually solve a problem by applying the different tools and ideas of mathematics, such as equations, functions, figures, probability, statistics and so on?
d Do you think it is important to study mathematics?

The answer to every question above was chosen from the following choices: (A) Very much; (B) A little; (C) Not so much; (D) Never. The results were as follows.

a *Relationship to reality*
The study shows that 43% of the students answered A or B in the pre-test and 61% in the post-test. In this case, we note a remarkable change in the students' attitude towards mathematics. From this result, we can deduce that Mathematics of Track and Field Events is one of the resources that can be used in mathematics education from which the students can learn the real usefulness of mathematics.

b *Necessity of actual activities*
The result reveals that 86% of the students chose A or B in both tests. The percentage sum of desirable changes was 23%, which indicates that their performances in these examinations are not so significant. This high-percentage result probably shows that the students have been exposed before to the different types of activity.

c *Application of the different mathematical tools*
It was recorded that 80% and 82% of the total students selected A or B in the pre- and post-test respectively. Comparing these data to those obtained in the study of Mathematics of a High Building (Yanagimoto 1991), where the students obtained 38% and 57% in the pre- and post-tests respectively, indicates

that the students were well equipped and aware of the different ideas of mathematics. However, the percentage sum of desirable changes was 24%, which does not seem to reveal any significant result.

d *Importance of learning mathematics*
The result here shows that 84% of the students chose A or B in the pre-test, while 86% did so in the post-test. The desirable changes were recorded at 25%, which implies that some changes have been made regarding the students' recognition of the merit and importance of studying mathematics. This might suggest that the students have realised that the main objective of learning mathematics is for daily-life activities, rather than merely taking the entrance examinations.

Students' interest in the lesson

In this section, we are going to discuss the result of the students' examination regarding their interest in each unit of the lesson. They were asked to choose their answers from the same list of alternatives as previously.

Figure 6 shows that the level of the students' interest in the lesson increased as they worked through the three Units. This implies that the materials and problems used in this approach are suitable to enhance students' interest. In spite of a complicated method in Unit III, students have shown their strong enthusiasm and interest, presumably because the problem was to improve their own relay performance.

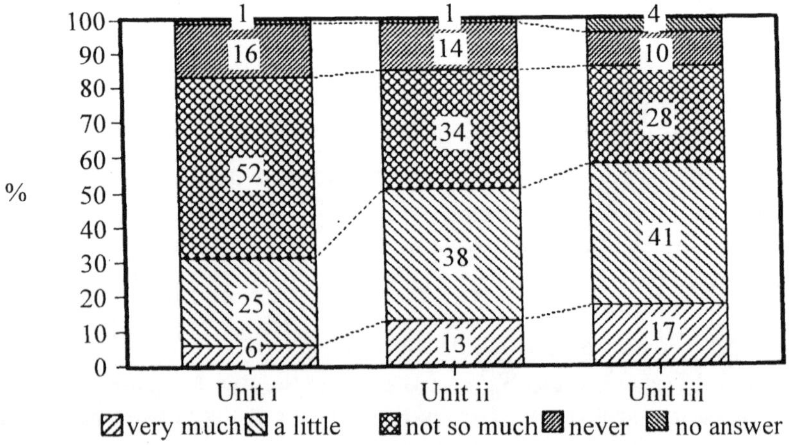

Figure 6. Diagram of students' interest in each lesson

5 CONCLUSION

The Mathematics of Track and Field Events was applied to demonstrate the usefulness of the CRM approach to third-year students of lower-secondary school. The results of the evaluation study can be summarised as follows.

- *Appropriate choice of materials.* The results show that the majority of the students were interested in Unit III, followed by Unit II. However, we have to consider the possibility that the degree of interest is dependent on the capability of the students to solve or tackle the problems by themselves.

- *Usefulness of computer.* By using the teacher-developed software, it was possible for the students to solve problems such as designing a race track. However the computer program should not be used by the students without understanding the mathematical theory behind it.

- *Evaluation of students' performance.* Through the CRM approach, the students became aware of applications of mathematics in the outside world, as well as the advantages of learning it. However, regarding the necessity of actual performance of activity, the application of different mathematical tools did not reveal a significant result.

From these, we can conclude that the Mathematics of Track and Field Event is a desirable and appropriate material for the CRM approach. However, further investigation of the CRM approach is recommended, by developing other materials and criteria, in order to have more conclusive results.

REFERENCES

Blum, W (1991). Applications and modelling in mathematics teaching - a review of arguments and instructional aspects. In Niss, M, Blum, W and Huntley, I (ed), *Teaching of Mathematical Modelling and Applications*. Chichester: Ellis Horwood, 10-29.

Blum, W and Niss, M (1989). Mathematical problem solving, modelling, applications, and links to other subjects. In Blum, W, Niss, M and Huntley, I (ed), *Modelling and Applications and Applied Problem Solving*. Chichester: Ellis Horwood, 1-21.

Kaiser-Messmer, G (1991). Application-orientated mathematics teaching: a survey of the theoretical debate. In Niss, M, Blum, W and Huntley, I (ed), *Teaching of Mathematical Modelling and Applications*. Chichester: Ellis Horwood, 83-91.

Matsumiya, T (1991). The purpose of mathematics education in Asian countries - in the case of Japan. In Yan, S *et al* (ed), *Proceedings of ICMI - China Regional Conference on Mathematical Education,* Beijing, 3.54-3.58.

Matsumiya, T, Yanagimoto, A and Mori, Y (1989) Mathematics of a lake - problem solving in the real world. In Blum, W, Niss, M and Huntley, I (ed), *Modelling, Applications and Applied Problem Solving.* Chichester: Ellis Horwood, 87-97.

Matsumiya, T, Yanagimoto, A and Mori, Y (1992). Mathematics of a railway. In Yokochi, K, Miyake, A and Watanabe, S (ed), Mathematics Education in Japan/For Five Nations Conference. New Jersey, 43-47.

Niss, M (1989) Aims and scope of application and modelling in mathematics curricula. In Blum, W *et al* (ed), *Applications and Modelling in Learning and Teaching Mathematics.* Chichester: Ellis Horwood, 22-31.

Yanagimoto, A (1991). Mathematics of a high building - problem solving in the real world. In Yan *et al* (ed), *Proceedings of ICMI - China Regional Conference on Mathematical Education,* Beijing, 3.66-3.71.

12

Units of mathematics in integrated projects with science and technology

Torben Steeg and Julian Williams
University of Manchester, UK

SUMMARY

This paper reports on findings based on experience and evaluation by teachers of over a dozen interdisciplinary projects in which mathematics units were integrated into real problem-solving projects in science and technology. These projects, and the kind of mathematics they involve, are described in general terms. It was found that the projects can benefit the motivation of students, and it is suggested that their learning of mathematics and of problem-solving strategies generally was enhanced. Furthermore the teachers perceived the experience to be a substantial professional development and inservice education for themselves.

1 BACKGROUND

This paper describes and generalises from the evaluation of interdisciplinary mathematics, science and technology projects with 11-16 year old students in over a dozen British secondary schools. In particular it draws on the oral and written evaluation reports of a group of eight schools, six of which have been published in McLachlan and Williams (1992). This report also incorporates a summary of the oral evaluation of the projects, conducted at a conference of the teachers involved.

In each case the project was the result of interdisciplinary collaboration by teachers of the three subjects; aims from the National Curriculum for the three subjects were identified, in advance, by the three departments. The students were involved in some projects which had 'real' outcomes, such as the making of some articles of clothing or the improvement of the school environment.

Another common feature was the large blocks of school time devoted to the project, ranging from several weeks up to a whole term. The management of the exercise was handled in very different ways in different schools, and there were many problematic issues for school organisation and management, which are published in McLachlan and Williams (1992).

2 AN EXAMPLE OF AN INTEGRATED PROJECT CONTAINING TWO MATHEMATICAL UNITS

In the early part of 1992, a whole year group of 12-13 year olds in a comprehensive school in Stockport conducted a project to improve their school environment. The brief given to the students was to design and build an area for student recreation and outdoor eating; it took place during their mathematics, science and technology lessons over a period of about eight weeks. The contributions of mathematics to this project were various, but two strongly mathematical tasks were identified – the site surveys and the anthropometric survey, (a survey of the dimensions of the students who were to use the outdoor furniture for their lunches). These units were developed by the mathematics department, though the students and staff understood that the purpose of the work was to contribute to the design and making of the final artifact; in this sense technology was the 'leader'. A unit involving an opinion survey was also involved, but we omit details of this here.

Site surveys

The school staff identified seven sites in the school grounds that were particularly suitable for improvement. Each of these sites was surveyed by a group of students, to establish such things as the overall dimensions of the site, its profile, the position of the site relative to the main school buildings, the access routes to the site, and the position of key existing features of the site including areas of shade and exposure.

The results from these surveys were used to produce accurate scale drawings of the sites, which could then be used in the design and making stages of the project. This required the students to measure accurately over relatively-long distances, and to scale their measurements appropriately. The students understood that the plans would be important to their design work, and that any faults might lead to disaster later. The diagram on Fig. 1 shows the site plan produced by one group of students as a result of the survey.

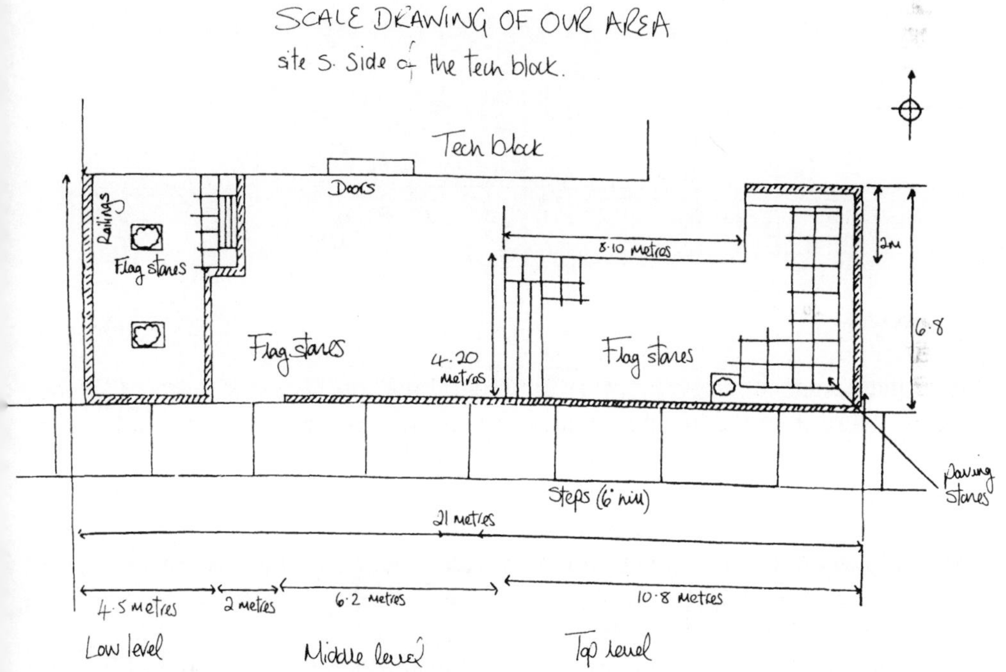

Figure 1

Anthropometric survey

One of the aims of the environmental improvement was to provide outdoor furniture for student recreation and eating. This lead to the need for data on the physical characteristics of the school population, which was collected and analysed so that the furniture could be designed with the most appropriate dimensions. Again, the information gained in the survey was processed to provide information for the design and making stages of the project.

The critical analytical task for the students in this context was to extract, from the large amount of data that they had gathered, information that could inform their design (see Fig. 2).

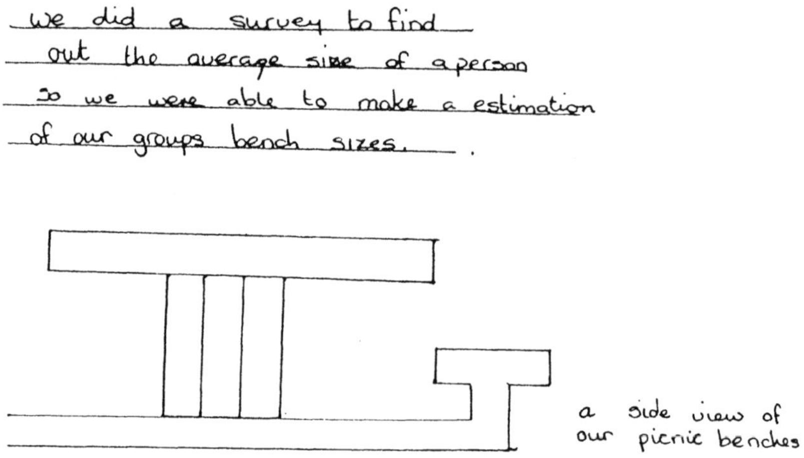

Figure 2

These mathematical activities were an integral part of the wider project that the students were engaged in, and the project itself was perceived by the students to be related to a real need within the school. Thus the students were tackling well-defined mathematical tasks, these tasks supported the learning and using of specific parts of the mathematics curriculum, and the students were highly motivated to do the tasks and to do them well, because of the relationship of the tasks to the wider project.

The embedding of the mathematical tasks in a well-defined and real project also led, we believe, to a better understanding of the mathematics by the students. For example, in the anthropometric survey the students gathered data from a proportion of the school population on weight, knee height and knee-to-back distance. The correct processing of these data, to extract useful design criteria for such things as picnic tables was critical to the success of the final artifact. The following extract from the students' work (Fig. 3) shows a part of this analysis, the extraction of a mean from a set of ranked data, linked to a description of the process used. Thus students were engaged in tasks such as finding averages and producing frequency tables, where the need for the data processing was clear and any mistakes in the processing would be likely to be noticed simply because of the context of the data. The context leads to a natural validation through a common-sense understanding of the real situation.

The Frequency of peoples weight, knee height and knee/back

WEIGHT	Total frequency	Knee Height	Total frequency	Knee/Back	Total Frequency
20 - 29	0	30 - 34	2	30-34	1
30 - 39	16	35 - 39	3	35-39	6
40 - 49	42	40 - 44	3	40-44	29
50 - 59	42	45 - 49	40	45-49	28
60 - 69	27	50 - 54	36	50-54	35
70 - 79	8	55 - 59	22	55-59	27
80 - 89	0	60 - 64	2	60 - 64	8
90 - 99	1			65 - 69.	2
	Total 136		Total 136		Total 136

To find mean from a grouped frequency table.

Weights multiply the mid group value in each group by the frequency

Weight	mid-value	frequency	mid value and frequency.
30 - 39	34 · 5	16	552
40 - 49	44 · 5	42	1869
50 - 59	54 · 5	42	2289
60 - 69	64 · 5	27	1741·5
70 - 79	74 · 5	8	596
80 - 89	84 · 5	0	0
90 - 99	94	1	94·5
		136	7142

Estimated mean = $\frac{7142}{136}$ = 52.51 kg.

Figure 3

It is not uncommon for students to fail to notice errors in everyday calculations, even when they lead to an incorrect order of magnitude. The design task here increases the likelihood that the student checks the answer for reasonableness, and is thus more aware of the need to round to an appropriate level of significance. On the other hand problems do arise, and opportunities can be missed. For instance, we expected a discussion of the range of the data, because of its importance in the design. We suspect that the reason for its omission lies in the mismatch of expectations between the mathematics teacher (who would regard such calculations as more appropriate to older students), and the technology teacher (for whom they are important because of the implications for the design).

3 GENERIC MATHEMATICAL TASKS

The above examples are typical of a range of tasks which are strongly mathematical, and which are commonly found in school technology projects. In Williams (1993) several such generic tasks were described, including the following.

The calibration of an instrument or device

Calibration arises regularly in both scientific and technological project work, and the mathematics required is well defined. However, science and technology teachers often take the mathematical know-how for granted.

Typical projects that will lead to a need for calibration include the design and making of a wide range of measuring instruments, and any artifact that requires setting to a particular value. In Steeg and Williams (1991) the calibration of a ballista was described. This required students to explore, practically, the relationships between such things as the angle of projection, and the height and length of the projectile path. Since the device was evaluated in terms of its ability to hit a target, it was necessary for the students to focus on the accuracy both of their collected data and the use of the data in the calibration.

Making a two- or three-dimensional scale model

This type of activity arose in many school projects. Such models are commonly constructed as design aids during the development stages of a project. A typical mathematics project is to plan a kitchen, bedroom or bookcase. The integration into project work allows such work to be situated in real projects, such as those described in McLachlan and Williams (1992).

A result of this reality is that there tend to be more constraints put onto the mathematical activity – for instance when planning a book-case, the wood grain and wastage from the sawing will have to be considered. A model ergonome will need to reflect the real antropometric data gathered.

The appropriate display and interpretation of quantitative data

Data can result from survey research such as that above, or of opinion surveys, or from 'scientific' investigations of the relationship between variables in a cause-and-effect model. It has been argued that good practice in technology requires a quantitative and scientific foundation, and of course this is highly motivating for mathematics. The 'appropriateness' of various techniques of display and interpretation can only be learned and can only be judged in the context of tasks which provide for an audience, such as the 'clients' who provided the design brief. The more 'real' those clients are, the more genuine the motivation for communication, and hence the richer the interpretation and display will be.

4 COMPARISON WITH TRADITIONAL CLASSROOM MATHEMATICS

There are a number of distinguishing characteristics of such mathematical activities when they are integrated into work in science and technology, which contrast with much mathematics conducted in traditional academic settings. In integrated work the mathematics is learned and applied in a well-defined context, located outside the mathematics classroom and, in many cases, outside the school.

In contrast the traditional approach has been that the mathematics is learned in the abstract, in a purely academic setting. It is assumed traditionally that such academic knowledge can then be applied to a real situation. However, there is now plenty of evidence that casts doubt on this assumption.

When mathematical tasks are located within wider technological or scientific activities, they are usually 'big' – normally taking from half an hour to several hours to complete – and often involving work in a variety of geographical and social settings. (An example is where data is collected by writing to local companies, and enquiries may be needed by telephone; work is often completed in workshops or outside in the locality.) In contrast, tasks in traditional mathematics curricula are short – maybe taking a few minutes or even less to complete – and are confined to the classroom, or even to the desk.

In integrated projects, the mathematical tasks are part of a project motivated by the context, and a perceived need or opportunity. By contrast, it is more usual for traditional tasks that students are engaged in to be motivated by the teacher, by the needs of an examination, or perhaps by their intrinsic interest.

Since the teaching is integrated across subjects, the mathematics teacher may act as one of a number of consultants to a broad spread of projects, where the activities are focussed on a technology workshop or a science laboratory. This contrasts strongly, perhaps threateningly, with the more common situation where the teaching is by a mathematics teacher, in mathematics time, in a mathematics classroom. The locus of control is inevitably, and irretrievably, shifted towards the student, and away from the teacher.

5 FINDINGS FROM THE INTERDISCIPLINARY PROJECTS

The following are findings that have been reported by the teachers involved in the planning and teaching of the projects. In several cases their evaluation involved an explicit evaluation by the students; in some cases we have interview and questionnaire evidence from the students themselves. There is clear evidence that motivation of mathematical work is enhanced. Many teachers also report signs of improved mathematical understanding, though there is a case for suspending judgment on this in the absence of rigorous evaluations of performance. Another strong outcome, reported by teachers involved, is in their own professional development and inservice training.

Students' *understanding and learning* of mathematics, and problem-solving skill generally, is enhanced. According to reports by the teachers involved, this may occur in a variety of ways.

- By *removing conflicts, inconsistencies and repetition*, for example between the methods used by different subject teachers to solve an equation, or the different names given to the same concept ('nets' and 'developments', for instance). Avoidance of repetition (either for pupils or staff) may occur, for example, when measurement or data-analysis skills that are common to all three subjects might be taught and assessed once only. Alternatively one piece of coursework can assess students' understanding of the use of measuring instruments, and their ability to use them accurately, as required by mathematics, science and technology curricula, (Maughan *et al,* 1990).

- By *drawing on the problem-solving and investigative aims* of the curriculum for science and technology. For example, students taught strategic skills of evaluation in technology can be encouraged to make connections with 'looking back' strategies in mathematics. Similarly, the communication and leadership skills involved in working together in teams, and reporting to an outside agency, can help the students throughout the curriculum and in their general education.

- As the *increased mathematical and scientific rigour* in project work leads to a better understanding of how mathematics can be *properly* applied in real problems.

Pupils' *motivation* for doing mathematics, as part of the project, is also enhanced.

- By *linking the academic knowledge with 'doing'*, such as making. This can be particularly motivating where a student who dislikes mathematics has to succeed in a mathematical task for the sake of the wider project.

- By *working with a real problem of personal or social importance*. This can be motivated by industrialists and engineers, who provide real problems from outside the school environment, but commonly is motivated by the needs of the pupils themselves in their own lives. In several cases the projects have been inspired by the needs of local primary children (such as The Young Child project in McLachlan and Williams, 1992) or the community.

- Through *contact with outside agencies* for help and information, for example when collecting preparatory research data, and through public display or presentation of mathematical work that has arisen from the project. By working with people (pupils, staff and other adults) as equals, the students' self esteem may be enhanced.

- Through *use of larger blocks of time than usual*, thus preventing the loss of momentum that is often caused by breaking off an activity at the end of a lesson, and also providing the students with experience of adult working conditions.

- The teachers' *professional development* and education were enriched.

  Just as the students were motivated by their involvement in communication with others with status, we found this was mirrored by the teachers' own reports of raised self-esteem. They clearly enjoyed their rôle in managing a development with teachers from other disciplines, and with adults from outside the school, including industrialists and parents.

  They also learned from the various different working conditions and teaching styles predominant in other subjects. For instance, mathematicians were exposed to the teaching styles of the laboratory and the workshop and they were introduced to the different modes of delivery and assessment involved in craft work.

6 CONCLUSIONS

We have discussed the kinds of mathematical activity which typically arise in integrated mathematics, science and technology project work, based on evaluating such projects between 1988 and 1992, giving examples which cover a large part of the curriculum for mathematics. Typically, these strongly-mathematical activities are situated in practical contexts, motivated by need, large in 'scope' and time, and student centred.

We find particular benefits for the students' enhanced *understanding* of mathematics arising from these projects because of

- the removal of inconsistencies and repetition between the teaching of mathematics, science and technology,
- synergy between the problem-solving processes taught in mathematics, science and technology,
- the appropriateness of the mathematical applications.

We find benefits for the students' *motivation* for learning mathematics through such project work, arising from the linkage with

- other subjects, their teachers and methods,
- alternative ways of working,
- other adults from outside the school.

A major impact on motivation arises from their association with the needs of an end-user.

Finally, we find benefits for the education and professional development of the teachers involved.

REFERENCES

Cherouvim, N *et al* (1991). *Practical Projects with Mathematics.* Cambridge: Cambridge University Press.

Dallow, R *et al* (1991). Mathematics, science and tchnology: springy things inc. In *Mathematics Teaching*, **135**, 7-10.

Kitchen, A *et al* (1992). *Made to Measure.* Century Maths. Cheltenham: Stanley Thornes.

Maughan, C, Steeg, T and Williams, J S (1990). *Bouncing Balls.* Manchester: University of Manchester, Department of Education and (in press) in *School Science Review.*

Mechanics in Action Project (1991). *Mathematics, Science and Technology in Schools.* Cross-Curricular Conference Report. Manchester: University of Manchester, Department of Education.

McLachlan, M and Williams, J S (ed) (1992). *Collaborative Work between Maths, Science, Technology in Stockport Schools.* Stockport: Stockport Local Education Authority.

Steeg, T J and Williams, J S (1991). Making ballistas with mathematics, science and technology. In *Design & Technology Teaching*, **23**, 152-1.

Williams, J S (1993). Linking mathematics, science and technology in UK secondary schools. In de Lange, J *et al* (ed), *Innovation in Maths Education by Modelling and Applications.* Chichester: Ellis Horwood, 365-374.

13

Implementing and assessing mathematical modelling in the academic 16-19 curriculum

Ann Kitchen and Julian Williams
Mechanics in Action Project, University of Manchester, UK

SUMMARY

The five-year pilot scheme for a new academic curriculum and examination in mathematics, that incorporated modelling and problem solving at its heart, has just been completed successfully. The scheme, for students between the ages of 16 and 19 years, showed that, as well as introducing modelling to the upper-secondary curriculum, the course widened access and improved the rate of student success. The course and examination is now being widely taken up throughout England and Wales; an estimated 16 000 students started the course in September 1992. The paper outlines the changes in teaching methods and assessment and, consequently, issues of teacher education and dissemination necessary to implement modelling and problem solving in the mainstream curriculum.

1 INTRODUCTION: WHY A NEW ADVANCED LEVEL?

Mathematics teaching for 16–19 year olds in England and Wales

Approximately 12.5% of the population complete a two-year course for the academic examination in mathematics (so-called A-level) at the age of 18. Of these about 25% fail the examination (Table 1).

	<—		Pass		—>	<—Fail	—>
Grade	A	B	C	D	E	N	U
1991 (n = 74,972)	19%	15%	14%	14%	13%	10%	15%
1990 (n = 79,747)	17%	15%	14%	14%	13%	11%	16%

Pass rate 1991: 75% (1990: 73%)

Table 1. Percentage of students achieving each grade at A level mathematics. Source: School Examination and Assessment Council.

The figures in the table underestimate the problem, since many of those starting the course at age 16 dropped out before the final examination and were therefore excluded from the data.

Mathematics has been declining in popularity for many reasons. These may include the following perceptions of students and teachers:

- mathematics is too difficult and many students fail the exam at the end of the course,
- mathematics is dull and boring,
- mathematics is too abstract and is not relevant to life.

It was obvious therefore that a new course was needed to improve students' motivation, and hence their success rate, as well as widening access to all those students within the top 35% of the ability range. This course should include modelling and problem solving to make it more interesting and relevant. In this paper we describe the work of the Mechanics in Action Project (MAP) in one such course, 16-19 Mathematics with Applications (Dolan, 1990).

The pilot scheme

A small-scale pilot, involving 32 schools, was set up by the School Mathematics Project (SMP), with the support of the Mechanics in Action Project and others, to develop a curriculum for A-level which included the teaching of mathematical applications and modelling among its aims. The importance of modelling in mathematics has been recognised for many years. However until recently, in England and Wales, its teaching had been restricted to a few classrooms. Various bodies such as MAP had worked with small groups of teachers to disseminate the work and to encourage good practice, but it had not had a widescale impact. It became apparent that only by changing the curriculum, and its means of assessment, for one of our national examinations would the necessary expansion take place. The results from the pilot study showed clearly that, as well as introducing successfully the teaching of modelling and problem solving to the 16 - 19 age group, the course improved motivation and widened access, while reducing the failure rate. Table 2 shows the improved pass rate among the 1991 cohort of students.

A level Grade	A	B	C	D	E	N	U
1991 (n = 849)	20%	15%	21%	17%	14%	8%	4%

← Pass rate 88% →

Table 2. SMP 16-19 Mathematics pilot schools' results by grade achieved (32 schools).

The dropout rate among the students was reported by schools to be less than half the normal dropout rate. This was despite the fact that the schools had widened access to those with lower grades on entry than previously. Schools reported that the changed assessment and the inclusion of modelling and problem solving, encouraged improved motivation and enjoyment. Thus it appeared to be possible to introduce innovation, such as modelling and coursework at the same time as widening access.

2 THE TEACHING AND ASSESSMENT OF MODELLING

In order to change both student and teacher behaviour, modelling and problem solving must be assessed. Otherwise it would be recognised by both teachers and students as a worthy part of the curriculum, but then ignored. This was recognised within the new syllabus, produced in cooperation with the Joint Matriculation Board (JMB), now part of the Northern Examination and Assessment Board, one of our national assessment bodies. The new course has assessed the 'important', rather than restricted the teaching to that which is easy to assess. Among the aims were the following (JMB, 1991). "To develop skills of modelling, ... to develop an appreciation of the links between mathematics and the real world ... to be able to model both in familiar and unfamiliar contexts ...". The objectives – those things which should be assessed – included "the ability to identify problems ... to set up a model ... to translate a mathematical solution into everyday language ... to validate results".

It was been recognised that a timed terminal-examination could not assess these skills. In addition a single examination had proved unsuitable for motivating many students throughout the course. The assessment was therefore widened to include instruments more suitable for mathematical applications (Applied Units), while keeping the conventional methods of assessment for pure mathematics. The content of the Applied Units is drawn from mechanics and statistics and other mathematical applications. That for the Pure Units is based on algebra, calculus, mathematical methods, numerical analysis and trigonometry. The method of assessment (JMB, 1991) arrived at in the light of the pilot scheme, was therefore as follows.

Assessed coursework marked during the course (Applied Units)	23%
Tests set and marked during the course (Applied Units)	27%
Final examination (Pure Units)	50%

Table 3.

The first two of these helped modelling and problem solving to be assessed in a meaningful way. 'Coursework' is defined as work done as an integral part of the course. It is not done under timed test-conditions, the students being free to discuss the problems and work at their own speed. The assessment is based on a written report by the student, which must be the student's own, unaided work. The assessed coursework varies from a short two-hour investigation of a given kinematics problem to an eight-hour project on a mathematical topic of the student's own choice. In all, nine pieces of work are assessed. This coursework can be used to assess the whole modelling process and to explore a real context. In other words to assess the ability to

> define a problem, analyse it, interpret the solution, validate results (through data collection or observation), demonstrate appreciation of model refinement, and demonstrate planning and communication skills.
>
> (JMB, 1991)

However, many difficulties became apparent during the pilot scheme. These can be summarised as follows.

- Modelling was unfamiliar to many teachers.
- Coursework assessment was unfamiliar to many teachers.
- It was necessary to achieve, and demonstrate publicly, a common standard nationwide.
- The assessment itself was found, in some schools, to detract from the learning process.

3 OVERCOMING TEACHERS' DIFFICULTIES WITH MODELLING

It was apparent during the trialling period that many teachers were unhappy about teaching modelling, an area of mathematics in which they had had no formal training. This was overcome by providing structured teaching-materials such as the unit guide and student text for the unit Newton's Laws of Motion (16-19 Mathematics, 1991a, 1991b). These included both teacher and student materials which encouraged the learning of modelling skills. For the first time both students and teachers were encouraged to look at the modelling process as an integrated part of their mathematics that spread through mechanics and statistics. For example, the initial mechanics unit included a practical investigation into kinematics that was built on throughout the unit. A kit of apparatus was developed to enable the students to collect their data quickly and efficiently (Unilab, 1990). The students learned to build a mathematical model and, subsequently, a Newtonian model for the motion, and write a report for assessment. Later units, for example Modelling with Force and Motion (16-19 Mathematics, 1992), teach students to develop longer modelling projects, of about 8 hours duration, in which they are expected to identify their own problem and develop appropriate practical and investigational work to solve it. An example might be to model the motion of a fairground ride (Kitchen, 1991; Williams, 1990).

In addition, in-service training (INSET) courses were held to enable teachers to experience this form of learning themselves. Feedback from the course evaluations showed clearly that the teachers valued these sessions of doing and investigating more highly than sessions where a more didactic approach was used. Some found that, by putting their own theoretical knowledge in a real context, their own misconceptions were challenged and corrected. Those attending were encouraged to work with others in their schools, who had not attended a course, to disseminate these skills, and INSET materials were produced to aid them in this work (SMP, 1991; MAP, 1991, 1992). An INSET video has been made illustrating how students might tackle an expected project on fairground ride (Savage and Williams, 1992).

4 THE ASSESSMENT TASKS

The A-level course was structured so as to provide assessment materials which encourage the learning of modelling and problem-solving skills throughout the two years. This was done in several ways. The initial assessed tasks were carefully devised so that many different solutions are possible. Fig. 1 shows part of the tasksheet for one of the short kinematics investigations.

A group of canoeists on holiday at the seaside decide to have a race off-shore. They set up a triangular course using a buoy and two other floats, with the start and finish at the buoy. They have been told that the prevailing current flows parallel to the shore at a speed of about 2 m/s. If the total course is 300 metres long, investigate where they might put the other two floats.

How does the layout of the course affect the time taken to complete the course?

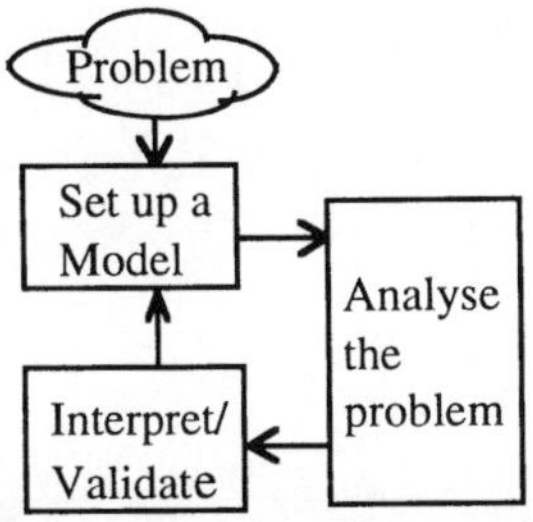

1. Set up a simple model to start.
   Make assumptions.
   Draw diagrams.
   Choose variables and assign constant values as appropriate.
2. Continue with the problem using your simple model.
   Interpret your solution.
3. Discuss the validity of your solution in the light of the limitations of your model
4. What changes or refinements could you make to your model?

Figure 1. A kinematic modelling task

Students are encouraged to set up a simple model for a part of the problem and to investigate that (see Fig. 2). They are not expected to find a perfect general solution to the problem.

Set up a model (Assumptions)

The beach is far enough away so that it does not affect the course shape.
The current is a constant 2 ms^{-1}.
The wind has negligible effect on the canoeists.
Negligible space is taken going round the buoys.
The course is an equilateral triangle with sides of 100 metres.

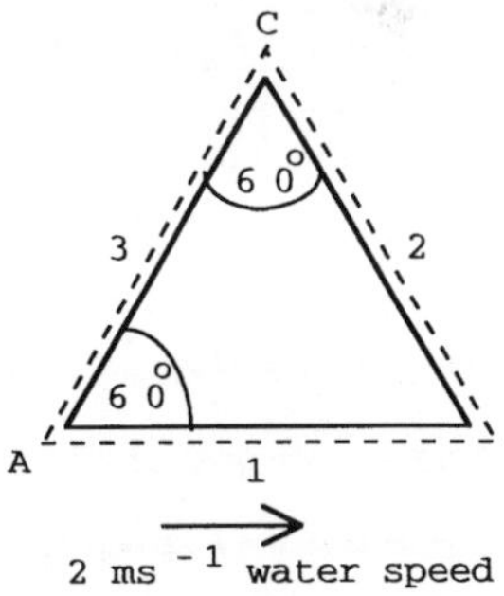

The rowing speed is a constant v ms^{-1}(in still water).
¿ is the angle, measured anticlockwise between the first side and the direction of the current.
t is the time, in seconds, taken to complete one lap, assuming that other canoeists do not affect the time.

Figure 2. Example of a student setting up a model for the above investigation

The initial work is done in class, and students are encouraged to discuss the problem with each other, after which they are expected to write a final report on the investigation. They are given guidelines on the assessment criteria their teacher will use (Fig. 5).

In the problem, some students assumed that the canoeing course set would be an equilateral triangle, and investigate the effect of rotating it relative to the direction of the current using column vectors. One student started by aligning the first leg of the course with the current, as in Fig. 2, and assuming that the paddling speed through the water was 5 ms^{-1}. Figure 3 shows part of her analysis.

The water velocity, v, can be broken down into two components

$$v = \begin{bmatrix} 2\cos 240 \\ 2\sin 240 \end{bmatrix}$$

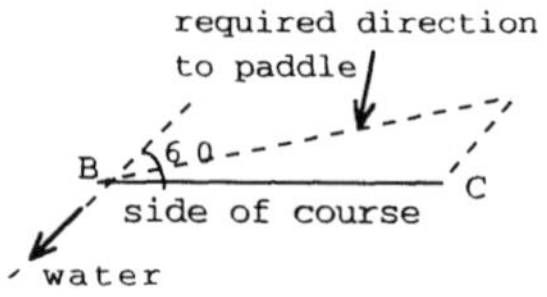

The required velocity of the canoe can also be split into components

$$\begin{bmatrix} m \\ n \end{bmatrix} \quad \text{where } \sqrt{(m^2 + n^2)} = 5$$

In order for the boat to go along the line BC the vertical component of the canoe(n) must equal the vertical component of the water but be in the opposite direction.
This implies that $n = -2\sin 240$ so $m = \sqrt{(25-(-2\sin 240)^2)}$

The resultant velocity in the y direction is zero so can be ignored.
The resultant velocity in the x direction is the x component of the canoe added to the x component of the water(2cos 240)

so the resultant velocity $= \sqrt{(25-(-2\sin 240)^2)} + (2\cos 240)$
$= 3.69$ ms^{-1} along BC.
The time taken for that side = 100/3.69 = 27.1 seconds

Figure 3. Sample of a student's analysis of the canoeing problem

Others decided to use isoceles triangles as the course, with the unequal side parallel to the current, and used scale drawing to see what happened as the sides changed in length, given a fixed canoe speed relative to the water (Fig. 4). These were only two of the wide variety of approaches to the problem used by students. This diversity means that students can discuss the problem, and work together as a group, and still be able to express their individuality in their own reports. Experience with the first cohort of 8 000 students has shown that the provision of these problems for coursework assessment has not led to the routinisation of the task into a set of fixed recipes.

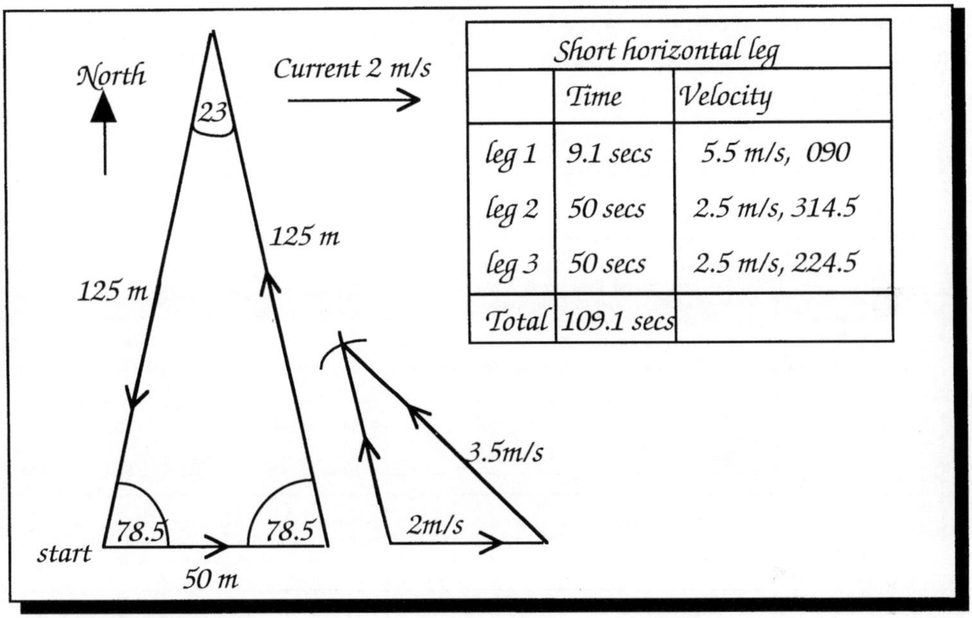

Short horizontal leg		
	Time	Velocity
leg 1	9.1 secs	5.5 m/s, 090
leg 2	50 secs	2.5 m/s, 314.5
leg 3	50 secs	2.5 m/s, 224.5
Total	109.1 secs	

Figure 4. Sample of a second student's analysis of the canoeing problem

One problem that has become apparent during the monitoring of the first cohort is that teachers are unsure of how much help to give students. Some tend to shy away from any involvement in the task at all. This is unfortunate. While the task has to be assessed for examination purposes it is also intended to be part of the learning process, and teachers are expected to give advice and help where needed. In order to enhance the formative nature of the assessment the students do two tasks for their early assessments, the best mark being chosen from each strand to make up their final mark. Teachers are expected to provide feedback for the students from the first of the two tasks before the students attempt the second, so that they may learn from any mistakes they make as well as gain reinforcement of the positive achievement shown. This is a vital part of the learning process.

5 USING THE ASSESSMENT GRID

The assessment grids (Fig. 5) have been devised to help the teachers to give credit to modelling skills without being unduly prescriptive. Very precise criteria may discourage creativity, since the student may become obsessed by the need to satisfy every criterion set out, and the investigation as a whole becomes stultified.

Date.................... Title of Investigation 1 x.............................. 2 o.............................. Total Mark

Model Formulation	Problem defined and understood. Appropriate choice of variables and some simplifying assumptions	Appropriate model set up. Simplifying assumptions clearly stated and discussed.
Data collection and analysis	Data collected and /or organised as appropriate. Accurate calculations and some analysis of the problem	Clear and logical use of appropriate numerical, graphical or algebraic analysis.
Interpretation/ validation	Predictions/inferences stated in everyday language. Some attempt to test predictions as appropriate.	A reasoned attempt to interpret and validate/justify the solution(s) and hence identify both limitations and necessary refinements.
Communication/ Initiative	Work clearly presented and organised with some explanation. Problem tackled with persistance (with some help and guidance if necessary)	Effective use of mathematical language and diagrams. Clear explanation of the findings and their impli cations for further investigation. Strong personal contributions and/or self motivated research.

0 5 10 15 20

Figure 5. The assessment grid for coursework in mechanics and statistics units. Source: JMB, 16-19 Mathematics, Instructions and Guidance, 1991.

On the other hand, the teacher must be given sufficient help so that the assessment can be consistent. The INSET courses mentioned above, where exemplar coursework is marked and discussed, have proved beneficial, as has the holding of informal moderation meetings for small groups of schools. On the whole teachers have found little difficulty agreeing the ranking of work, using the grid without any training. Interestingly enough, the problems in standardisation seem to fall into two categories. Firstly, while the rank order given by different teachers to a batch of scripts was usually the same, the actual marks given may vary for different markers. Once exemplars for full and half-marks had been given, and the reasons for the marks discussed in training, the teachers found it fairly easy to adjust their marks accordingly. This seemed to carry forward to the next task, so that once the teachers had an idea of the standard required they were able to mark appropriately. This process has been successful so far for the first unit. Trialling has shown, however, that more INSET work will be needed to help teachers to mark the more extended and more individual project work of later units consistently.

Secondly the ranking itself broke down, as well as the percentage awarded in the case where a student produced a report where the analysis was faulty at some stage – a significant proportion of teachers initially gave lower marks on both the interpretation/validation strand and the communication/initiative strand than were deemed to be justified by the moderators. Table 4 gives the mean marks for a script of this type (MAP, 1991), given by a group of 51 teachers at a coordination course. The script itself showed that the student was not able to understand the correct use of vectors. However the work showed some merit in building on the faulty analysis, and the moderators felt that the last two strands of interpretation/validation and communication/initiative should be marked by ignoring the error in analysis.

	Model formulation (20)	Data collection and analysis (20)	Interpretation/ validation (20)	Communication/ initiative (20)
Moderators' mean mark	12	8	13	12
Teachers' mean mark (n=51)	13	9	9	9

Table 4. Mean of marks awarded to the script, by strand.

This illustrates the importance and function of in-service training courses, and informal moderation meetings, if the marking is to be consistent across the country.

6 THE END-OF-UNIT TEST

As mentioned earlier, each of the Applied Units, in addition to coursework, is assessed by a one-hour test, set by the validating board and marked by the teachers using the board's mark scheme. The test would be used to assess syllabus coverage and mastery; it was felt that it should be possible to include some assessment of modelling skills within this test. Each test for the mechanics units has four sections, A to D. Three of these cover the conventional content, and the fourth, section D, is based on an article describing a real situation given to the students several days before, and is used to assess some modelling skills. A typical article might be about golf, describing the different sizes of ball, the layout of the course, the surfaces of greens, and the angle of flight produced by different clubs. They might then, in the test, be asked to set up a model of the motion of a ball across a fast green, to find the minimum speed with which a ball must leave the putter head so that it reaches the hole 15 metres away. The students would be expected to choose suitable constants, define variables, draw a diagram and so on (MAP, 1992). It was found that, in order for these questions to be useful the teachers and students needed exemplars of both questions and answers; these have been provided. This is sometimes thought to be undesirable as it might foster the production of rote answers, but this does not appear to be the case so far. The additional confidence engendered, especially where the teachers themselves are learning to model along with the students, more than outweighs any disadvantages.

The marks from 3 schools (a total of 62 students) for the first mechanics unit were collected. Each section of the test has equal weighting of 15 marks giving a total of 60 marks. The coursework is marked out of 40. Table 5 shows the mean marks gained.

Test	35.8	(out of 60)
Coursework	33.5	(out of 40)
Total – for 62 students	69.3	(%)

Table 5. Mean of students marks for *Newton's Laws of Motion* test and coursework

Initial statistical analysis of the marks supports the conclusion that the fourth section is assessing different skills from the first three, as we expected. The relatively higher marks on coursework are notable, and reflect the fact that students are given more time and help in the modelling process. An analysis of the coursework projects, and end-of-unit tests, suggests that, while the end-of-unit test is effective in assessing students' ability to understand assumptions within a given model, and to interpret a solution or discuss conclusions within a timed framework, the coursework is needed to assess students' ability to perform the modelling process as a whole, and appreciate the rôle of problem identification and model refinement therein.

7 CONCLUSIONS

The experience of the Mechanics in Action Project within the development of 16–19 Mathematics has shown that it is possible to implement a modelling curriculum into a mainstream A-level course, tackling problems of examination and teacher education, while widening student access and raising motivation and achievement. Key lessons from this experience concern:

- the provision of robust, trialled books, equipment and project ideas which integrate modelling into the learning of mathematics;
- the provision of a simple, clear, but flexible assessment scheme for modelling objectives involving timed tests and coursework;
- teacher training, through accreditation courses and moderation meetings, which involve the teachers themselves in modelling and in assessing exemplars of students' work on tests and projects against the course objectives.

We would claim that modelling and project work, combined with coursework assessment, has contributed positively to the improved motivation and success of the students with a new A-level course. In addition it appears to have widened the accessibility of the course.

REFERENCES

16-19 Mathematics (1991a). *Newton's Laws of Motion.* Unit Guide. Cambridge: Cambridge University Press.

16-19 Mathematics (1991b). *Newton's Laws of Motion.* Cambridge: Cambridge University Press.

16-19 Mathematics (1992). *Modelling with Force and Motion.* Cambridge: Cambridge University Press.

Dolan, S D (1990). 16-19 Mathematics. In *Teaching Mathematics and its Applications*, **9**.

Joint Matriculation Board (1991). *Syllabus for 16-19 Mathematics with Applications.* Manchester: Joint Matriculation Board.

Kitchen, A (1991). Alton towers: a practical approach to modelling. In Niss, M, Blum, W and Huntley, I (ed), *Teaching of Mathematical Modelling and Applications.* Chichester: Ellis Horwood, 260-270.

Mechanics in Action Project (1992). *Motivation, Organisation and Assessment of Coursework for the Optional Mechanics Units.* Manchester: University of Manchester, Department of Education.

Mechanics in Action Project (1991). *Organisation and Assessment of Coursework for 'Newton's Laws of Motion'.* Manchester: University of Manchester, Department of Education.

Savage, M D and Williams, J S (1992). *Modelling Circular Motion: an INSET Video.* Leeds: University of Leeds, Mechanics in Action Project.

School Mathematics Project (1991). *Organisation and Assessment of Coursework for 'Problem Solving'.* Southampton: SMP 16-19 Mathematics.

Unilab Ltd (1992). *The Practical Maths Kit Catalogue.* Unilab: Blackburn.

Williams, J S (1990). Modelling with mechanics 16-19. In Harding, D and Hirst, K (ed), *Mathematics 16-20.* Proceedings of the London Mathematical Society Under-graduate Teaching Conference. Nottingham: University of Nottingham, Shell Centre.

14

Iterative models in the secondary mathematics curriculum: some examples

Jonathan Choate
Groton School, Groton, USA

SUMMARY

Recent advances in technology have made it possible to introduce into the secondary mathematics curriculum a variety of iterative methods, which allow students to create simple models which can be used to study a variety of different types of dynamic behaviour. In addition, these methods can also be used to introduce students to one of the most exciting new areas in mathematics: chaotic dynamical systems. This paper will discuss these methods, and show how they are used in the mathematics curriculum at Groton School, a secondary school located in Massachusetts, USA.

1 DISCRETE MODELS

The availability of powerful, easy-to-use computational devices has made it possible to incorporate simple iterative models throughout the secondary mathematics curriculum. Using discrete recursive models, students can model a variety of phenomena such as population growth, financial accounts, and how the human body deals with drugs. This article will describe some of these models and suggest some related problems for students to study.

All the following models may be constructed using a spreadsheet, a graphing calculator, or a programming language such as Pascal, BASIC or Logo. They are all discrete, recursively-defined sequential models, which are easy to state yet which require a great deal of computation to solve. The basic idea behind the models is simple – if you know the initial state of a model, and if you know a rule for calculating the state of the model in the next generation in terms of its state in the current generation, then the state of the model can be calculated for all future generations. For example, consider the following discrete model of the human body's response to an ingested dose of 50 units of some drug over a period of 12 hours. If the body eliminates twenty five percent of the drug present in the system per hour, then the amount of drug still in the system at the end of the first hour, $D(1)$, will equal seventy five percent of $D(0)$, the amount at the beginning of the hour. Thus $D(1) = 0.75D(0) = 37.5$ units. Similarly the amount after two hours is, $D(2) = 0.75D(1) = 28.125$ units. Using a spreadsheet one can produce the table in Fig. 1, which gives the amount of the drug at the end of each hour. This model is defined by

$$D(0) = 50 \qquad D(n+1) = 0.75D(n) \tag{1}$$

where n is a non-negative integer. Equation (1) defines a linear recurrence relation.

Worksheet1

	A	B	C
1			
2	Spreadsheet Drug Dosage Model		
3			
4	Time	Amount of Drug in System	
5	0	50	
6	1	37.5	
7	2	28.125	
8	3	21.09375	
9	4	15.8203125	
10	5	11.8652344	
11	6	8.89892578	
12	7	6.67419434	
13	8	5.00564575	
14	9	3.75423431	
15	10	2.81567574	
16	11	2.1117568	
17	12	1.5838176	

Figure 1. Spreadsheet model of body's response to an ingested drug

Problem 1: Suppose the drug being modelled is not effective medically if the amount in the system drops below 5 units. How long before the drug is no longer effective, if an initial dose of 100 units is taken?

2 LINEAR AND QUADRATIC RECURRENCE RELATIONS

Recurrence relations are a powerful modelling tool available to students once they understand recursive definitions of sequences. A recurrence relation is said to be linear if it is of the form $B(n+1) = aB(n) + b$. In addition to modelling the body's response to drugs, recurrence relations can also be used to model the behaviour of financial accounts and population growth.

In order to model the behavior of an account which pays seven and a half per cent per annum with an initial balance of \$1500, for a 10 year period, one could use the recurrence relation

$$B(0) = 1\,500 \qquad B(n+1) = 1.075B(n) \tag{2}$$

and a spreadsheet, to produce the model shown in Fig. 2.

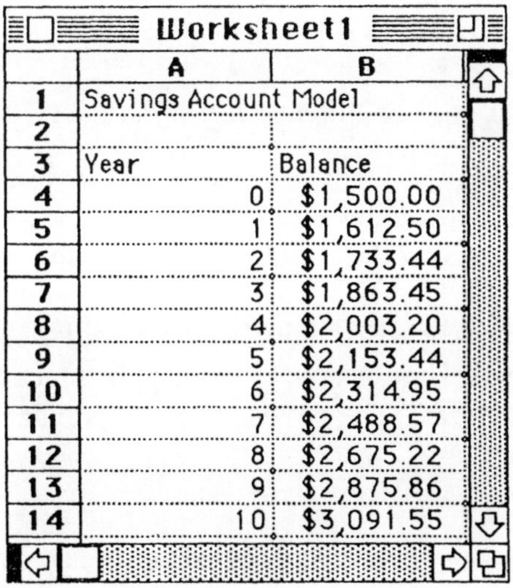

Worksheet1

	A	B
1	Savings Account Model	
2		
3	Year	Balance
4	0	$1,500.00
5	1	$1,612.50
6	2	$1,733.44
7	3	$1,863.45
8	4	$2,003.20
9	5	$2,153.44
10	6	$2,314.95
11	7	$2,488.57
12	8	$2,675.22
13	9	$2,875.86
14	10	$3,091.55

Figure 2. Spreadsheet model of a savings account.

Problem 2: How long will it take an investment of \$100 at your local bank to double in value? How much will it be worth in 20 years? In 50? In 100 years?

In order to convert this model into one where yearly deposits of \$250 were made, one could use the recurrence relation

$$B(0) = 1\,500 \qquad B(n+1) = 1.075B(n) + 250 \tag{3}$$

In general, an account paying r percent per interest period, with payments of P dollars, and an initial balance of b dollars, can be modelled using

$$B(0) = b \qquad B(n+1) = \left(1 + \frac{r}{100}\right)B(n) + P$$

If $b > 0$ the account is called an *annuity* and, if $b < 0$, a *loan*.

Problem 3: How much does it cost to finance your local megabucks lottery, if the payoff is \$50 000 per year for 20 years?

The type of growth exhibited by financial accounts is called *exponential growth.* Populations also grow exponentially, and can be modelled recursively. For example, a population with a birth rate of b percent per year, a death rate of d percent per year, E people emigrating per year, I people immigrating per year, and an initial population of p people, can be modelled using

$$P(0) = p \qquad P(n+1) = (1+g)P(n) + I - E \tag{4}$$

where g, the *net growth rate*, is equal to $\frac{b-d}{100}$.

Problem 4: Research the population growth figures for your country, and use the preceding model to predict the population for the next 10 years.

A major criticism of the preceding model is that populations usually do not have constant net growth rates. Here is how to modify the preceding model in order to use a variable net growth rate. The growth rate of a population during the $(n+1)$th time interval is given by

$$\frac{P(n+1) - P(n)}{P(n)} \tag{5}$$

If the growth rate is some constant, g, then

$$\frac{P(n+1) - P(n)}{P(n)} = g \tag{6}$$

Equation (6) is algebraically equivalent to

$$P(n+1) = (1+g)P(n) \tag{7}$$

which is the recursive model for exponential growth. For more on modelling with recurrence relations, see Maurer and Ralston (1991).

This model can be improved by assuming that the population lives in an environment which has a carrying capacity of M (that is, it can only sustain M people in the steady state) and that the growth rate of the population is some function of the ratio of the current population to the carrying capacity. If one further assumes that there is some maximum growth rate g that occurs when the population is very small, and that the net growth would be zero if the current population were close to M, then the net growth rate of the population during the $(n+1)$th time interval can be expressed as

$$g\left(\frac{M - P(n)}{M}\right) \tag{8}$$

Equating expressions (5) and (8) one gets

$$\frac{P(n+1) - P(n)}{P(n)} = g\left(\frac{M - P(n)}{M}\right) \tag{9}$$

which simplifies to

$$P(n+1) = (1+g)P(n) - \left(\frac{g}{M}\right)P(n)^2 \tag{10}$$

which is called the Logistic or Verhulst population model, which expresses $P(n+1)$ as a quadratic function of $P(n)$. In general, a population living in an environment with a carry capacity of M, with a maximum possible growth rate of g people per year, and an initial value of p, can be modelled using the recurrence relation

$$P(0) = p \qquad P(n+1) = (1+g)P(n) - \left(\frac{g}{M}\right)P(n)^2 \tag{11}$$

This model is used by wildlife biologists to help manage animal populations. They often have to decide how many members of a herd to kill off or to harvest. Here are two possible harvesting strategies.

• ***Plan A***: Harvest the same number, H, of animals per year. This can be modelled using

$$P(0) = p \qquad P(n+1) = (1+g)P(n) - \left(\frac{g}{M}\right)P(n)^2 - H \qquad (12)$$

• ***Plan B***: Harvest the same percentage, h, of the population per year. This can be modelled using

$$P(0) = p \qquad P(n+1) = (1+g)P(n) - \left(\frac{g}{M}\right)P(n)^2 - \left(\frac{h}{100}\right)P(n) \qquad (13)$$

Problem 5: Compare the two plans. If you were managing a wildlife population, which would you use and why?

The Verhulst model has recently received a lot of interest because it can be used as an example of chaotic behaviour. The following problem can be used to introduce students to the mathematics of deterministic chaos. For more on chaos and the logistic model, see Peitgen and Richter (1986).

Problem 6: Let $M = 1$, $p = 0.5$, and study the behaviour of the population defined by the recurrence relation

$$P(0) = 0.5 \qquad P(n+1) = (1+g)P(n) - gP(n)^2 \qquad (14)$$

for values of g such that $0 \le g \le 3$.

3 INTRODUCTION TO DYNAMICAL SYSTEMS USING ITERATED LINEAR AND QUADRATIC FUNCTIONS

All the models mentioned so far can also be developed using the concept of an iterated function. Iterated functions are an important concept in the field of dynamical systems, the study of processes that evolve in time. Here is how to iterate a function.

Select a function $f(x)$, pick a *seed* x_0, and calculate the sequence or *orbit* x_0, x_1, x_2, x_3, ... where $x_k = f(x_{k-1})$. Note that x_k can also be found by calculating $f^{[k]}(x_0)$ where $f^{[k]}$ denotes the composition of f with itself k times. Dynamical systems includes the study of the qualitative behaviour of orbits produced in this manner. An important concept in the study of orbits is that of a fixed point; $\boldsymbol{x}$ is said to be a fixed point for a function f if $f(\boldsymbol{x}) = \boldsymbol{x}$. Fixed points are one of the key concepts in dynamical systems theory.

The behaviour of the models discussed above can be understood using results from dynamical systems. The following theorems are useful and easy to prove.

Theorem 1: A linear function $f(x) = ax + b$ has a fixed point at $\boldsymbol{x} = \frac{b}{1-a}$ if and only if $a \neq 1$. If $a = 1$ and $b \neq 0$, then f has no fixed point.

Theorem 2: If $f(x) = ax + b$ has a fixed point $\boldsymbol{x}$, then f can be rewritten in fixed-point form as $f(x) = a(x - \boldsymbol{x}) + \boldsymbol{x}$.

Theorem 3: If f has a fixed point $\boldsymbol{x}$, then the kth iterate of f, $f^{[k]}$, can be expressed in the form

$$f^{[k]}(x) = a^k(x - \boldsymbol{x}) + \boldsymbol{x} \tag{15}$$

Since any model developed using a linear recurrence relation can also be expressed using iterated linear functions, the preceding three theorems can be used to study the qualitative behaviour of the models. Linear iterative models will either have orbits which get attracted to the fixed point, or they will have orbits which get repelled from the fixed points. The preceding theorems with appropriate restrictions also hold for f where

- $f(x) = (a+ib)x + (c+id)$, where a, b, c, and d are real numbers and i is the square root of -1.
- $f(X) = AX + B$, where A is a 2×2 matrix and X and B are 2×1 matrices.

Functions of the form $f(X)$ mentioned above can be used to develop much of the theory behind linear fractals, such as the Koch Snowflake curve and the Sierpinski Triangle shown in Fig. 3 which have recently become very popular.

Figure 3. Fractals created using iterated linear functions.
Left: Koch Snowflake curve. Right: Sierpinski triangle.

Unlike linear functions, the behaviour of quadratic functions under iteration is complex and often bizarre, and it is only within the past 15 years that it has been understood. The next two problems, along with problem 6, can be used to study the fascinating behaviour of the iterated quadratic.

Problem 7: What is known in the literature as the period-doubling route to chaos can be developed by studying the different orbits produced by iterating the function $f(x) = x^2 + c$. Start with a seed of 0, and see what happens to the orbits as c takes on values from 0.25 to -2.

Problem 8: Show that, if one iterates the complex function $f(Z) = Z^2 - 1.3$ starting with a seed of $Z_0 = x_0 + iy_0$ where $-2 \leq x_0, y_0 \leq 2$, the resulting orbit is either bounded or becomes infinite. The set of all seeds, Z_0, which produce orbits that are bounded is called the filled-in Julia set for $Z^2 - 1.3$. Next plot the filled-in Julia set for $f(Z) = Z^2 - 0.132 + 0.745i$ and compare it with what you got for $f(Z) = Z^2 - 1.3$. How are they different? How are they the same?

Problems 7 and 8 have kept many mathematicians busy over the past 25 years, and it is only within the last 10 years that complete answers have been found to both questions. They are both problems which can be understood by anyone with a secondary mathematics background, yet which have been studied in depth by research-level mathematicians. They are excellent for explaining to students what mathematicians do. For more on dynamical systems, see Devaney (1988).

4 MODELLING SYSTEMS USING SYSTEM DYNAMICS

System dynamics was developed by Jay Forrester (MIT) to model complex economic systems. The basic principle of this field is that, if you understand the structure of a system, you can understand its behaviour. One of the key concepts used is that of feedback loop, and systems are analysed in terms of the interaction of positive loops (those that create change) and negative loops (those that negate change). One of several programs which allows students to model systems using the principles of system dynamics is *STELLA* (Systems Thinking Educational Learning Laboratory with Animation). Unlike recurrence relations and iterated function models, STELLA allows the use of a variable time increment, so it may be used to create models which approximate continuous rather than sequential behaviour. STELLA provides an easy way for students to make the transition from discrete models to continuous ones. Put another way, it helps students to leave the world of pre-calculus and enter the world of calculus.

Models are produced in diagram form on the screen, using tank and valve icons. For example, the diagram for a simple savings account model is shown in Fig. 4, where the valve 'interest' controls the flow of money from the source into the 'Balance'.

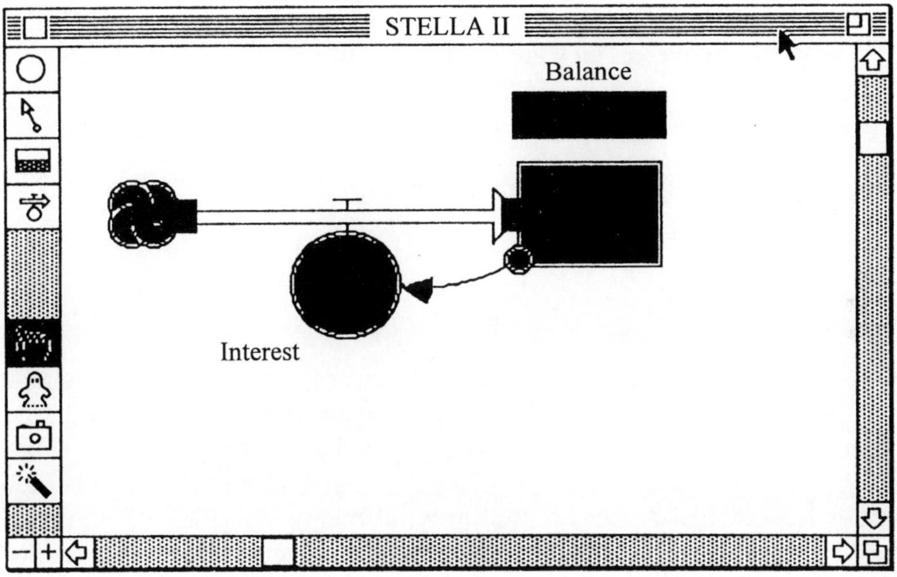

Figure 4. Diagram for a STELLA model of a savings account

Once the model is defined in diagram form, and appropriate initial values for the tanks and equations defining the valves are given, the program will automatically run the model giving output in both table and graphical form. STELLA makes it very easy for students both to define and probably more importantly play with a variety of models. The program allows the user to define parameters, and offers a sensitivity-analysis option which produces a series of runs of the model for differing values of the parameters selected. STELLA is very easy to use and makes it possible for students to produce sophisticated models of flu epidemics, prey predator relations, the AIDS epidemic, and the dynamics of the arms race.

STELLA uses numerical integration techniques, such as Newton's method, to produce models. These are all iterative methods and provide a natural extension of the work students have already done with iterated functions and recurrence relations.

One of the new calculus teaching programs, Calculus in Context, starts with a model of a flu epidemic. Students are given the following information, and then asked to predict the outcome of the epidemic.

- There are 3 populations that need to be considered: S, the susceptible population; I, the infected population; R, the recovered population. At the onset of this outbreak of the epidemic there were 45 400 people who were susceptible, 2 100 who were infected, and 2 500 who had recovered. Once a person has recovered, they are immune for life.

- The dynamics of the epidemic are given by the following rate equations

$$S' = -0.00001SI$$
$$I' = 0.00001SI - \frac{I}{14}$$
$$R' = \frac{I}{14}$$

where S' is the instantaneous rate of change of the susceptible population S, I' is the instantaneous rate of change of the infected population, and R' is the instantaneous rate of change of the recovered population.

Here is how STELLA can be used to build this model. First, three tanks are placed on the screen, and labelled Susceptible, Infected and Recovered. Next, the tanks are connected with two valves, which are labelled delta$S$ and delta$R$, which control the flow from one tank to another.

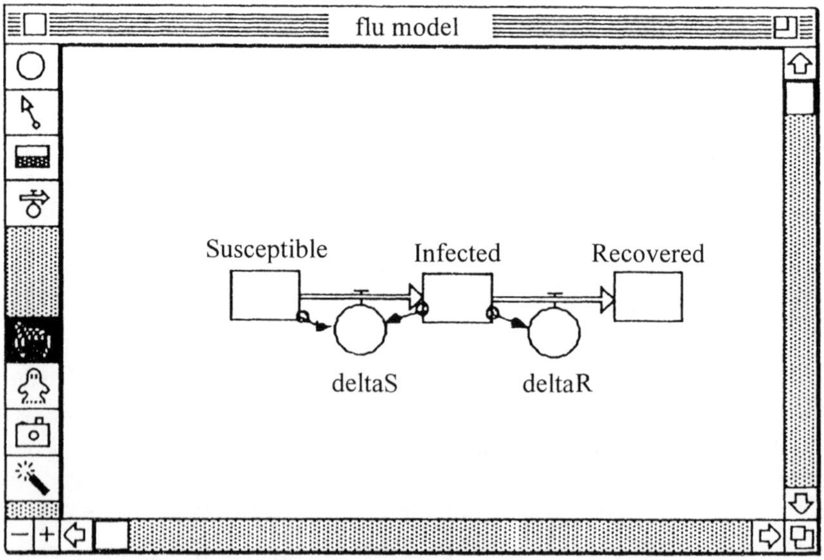

Figure 5. Diagram for a STELLA flu epidemic model

Since delta$S$ depends on both the susceptible population S and the infected population I, arrows are drawn from both the Susceptible tank and the Infected tank to the valve delta$S$. For a similar reason, there is an arrow from the Infected tank to the valve delta$R$. Since we have $S' = -0.00001SI$, there is an outflow from S to I, and the valve delta$S$ is set equal to $-0.00001SI$. Since $R' = \frac{I}{14}$, there is an outflow from I to R, and the valve delta$R$ equals $\frac{I}{14}$. Note that the way the model is set up – the net change in the infected population must equal $-(S'+R')$ so the equation for I' is implicitly incorporated in our model. The diagram for this model is shown in Fig. 5.

Running this model produces the graphs shown in Fig. 6. Once this model is produced students can start playing with the model, and try out different policies for dealing with epidemics. They can also change the model, to see what happens if the recovered population lose their immunity and become infected again.

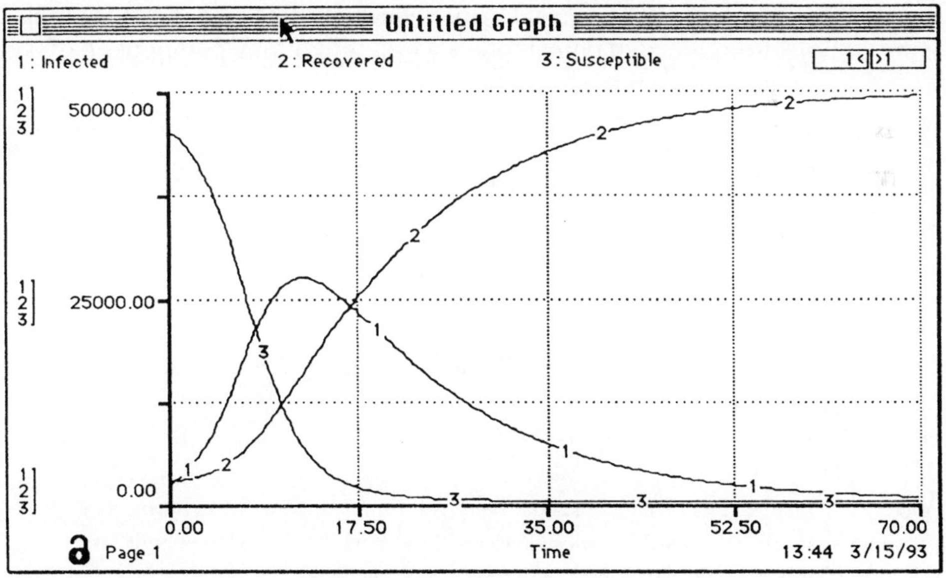

Figure 6. Graphs produced by a STELLA flu epidemic model

4 CONCLUSION

There now exist software packages such as spreadsheets and STELLA which make it very easy for students to model dynamic behaviour using iterative methods. These are appropriate for the secondary mathematics curriculum, and can be used to help students understand a variety of real-world phenomena. In addition, iteration is a topic rich in mathematical content, which can be studied on its own. Iterative models should become an important part of the secondary mathematics curriculum.

At Groton School we have integrated iterative methods into all levels of the curriculum, ranging from grade 8 to grade 12. They have allowed us to use dynamic modelling throughout our syllabi, and to show students a variety of interesting applications of mathematics. We use new technology extensively, and it has allowed us to create a curriculum which has better prepared our students for the increased quantitative demands both of college and of the job market. The feedback we get from recent graduates, and some of the colleges they have attended, has been very positive, and has convinced us that a modelling approach using simple iterative dynamic models is an important component of any secondary mathematics programme.

REFERENCES

Callahan, J (1992). *Calculus in Context. Five College Mathematics Project.* Northampton: Smith College.

Devaney, R (1988). *Chaos, Fractals and Dynamics: Computer Experiments in Mathematics.* Reading: Addison-Wesley.

Maurer, S and Ralston, A (1991). *Discrete Algorithmic Mathematics.* Reading: Addison-Wesley.

Peitgen, H-O and Richter, P H (1988). *The Beauty of Fractals.* New York: Springer-Verlag.

Sandefur, J (1990). *Discrete Dynamical Systems.* New York: Oxford University Press.

STELLA, High Performance Systems, 45 Lyme Road, Lebanon.

15

Mathematical modelling today and tomorrow

Robert Money
Camberwell High School, Victoria, Australia

SUMMARY

This paper uses case studies of the seven Australian state education authorities to illustrate progress and difficulties in incorporating mathematical modelling objectives into mainstream courses for a widening cohort of senior secondary students. Considerable progress has been made in specifying modelling objectives, in linking them to and weighting them against particular content skills. The framework within which different examining authorities share curriculum and assessment responsibility with classroom teachers are found to have important consequences for the quality of work done. Despite some successes, difficulties have been encountered in introducing modelling objectives into those courses which are more highly valued for university entrance. The paper makes some suggestions as to how these difficulties might be overcome.

1 BACKGROUND FOR CHANGE

Retention rates to the final year of secondary education have increased dramatically in Australia – from about 55% nationwide ten years ago, to a current 65% (still growing) in Western Australia, 82% in the more populous Victoria and 97% in the national capital, Canberra.

In some states, typically New South Wales, the debate for restructuring of senior secondary mathematics curriculum has been carried on mainly in the specialist community – of mathematics educators, tertiary mathematicians and mathematics users. In other states – notably South Australia, Victoria and Queensland – changes have taken place within a restructuring of the entire senior secondary framework. Examples are as follows.

- Mathematics, science or technology subjects are required in both Years 11 and 12 of the Victorian Certificate of Education (Victorian Curriculum and Assessment Board, 1990, 1992) – with the consequence that at least one mathematics subject is studied by 97% of Year 11 students and by 67% of final year (Year 12) students.

- Similar requirements in South Australia extend to a maths, science or technology component within a writing folio assessed in that state's certificate (Senior Secondary Assessment Board of South Australia, 1991, 1992).

National educational goals have been articulated through various reports and calls for a more skilled workforce. One report (Mayer, 1992) identifies 'Using Mathematical Ideas and Techniques' as one of seven key competencies, overlapping with others such as 'Expressing Ideas and Information and Solving Problems'. For the first time a National Statement on Mathematics in Australian Schools (Australian Education Council, 1990) has been formulated and national mathematics profiles, providing common levels for describing achievement, are being developed. Mathematical modelling, and other process objectives, are beginning to challenge content coverage – skills and techniques – in the debate over curriculum and assessment requirements at the secondary-tertiary interface.

2 MATHEMATICAL MODELLING: DEFINITIONS AND ISSUES

Other papers in this volume (Blum, Usiskin) give comprehensive definitions of the term 'mathematical modelling'. The draft Australian Mathematics Profiles (Australian Education Council) include – along with five content strands – a process strand entitled 'Working Mathematically'. Within this strand outcome statements for 'Using and Developing Mathematical Models' are described as follows.

- *Students in Level 6* (75% of Year 10 students)
  Choose between and use standard mathematical ideas and techniques to deal with situations where the solution method is not immediately clear.

- *Students in Level 7*
  Identify assumptions in standard methods and adapt mathematical models to provide better fits with situations.

- *Students in Level 8*
  Formulate a mathematical model by making useful and simple assumptions, collecting any data needed and representing the relevant relationships in mathematical terms.

While these high-sounding statements sit comfortably with Blum's inclusive definitions the current situation, as expressed in assessment requirements and student performance at senior secondary level, is less satisfactory. Two contradictory perceptions of applied mathematics, identified by Usiskin, have their Australian adherents.

- Applied mathematics is appropriate only for the best students – particularly if it involves calculus and mechanics.
- Applied mathematics is an appropriate alternative for those students who cannot do pure mathematics.

Efforts in the different states to implement mathematical modelling objectives more fully can be compared from a number of perspectives. Examples are

- the ways in which courses are structured for different sections of the student cohort;
- the extent to which modelling is valued by post-secondary users, in particular in first-year tertiary and teacher-education courses;
- the balance in emphasis between course content and process objectives;
- the extent to which assessment arrangements target modelling skills;
- the extent to which model-rich topics – such as statistical modelling, discrete mathematics and descriptive calculus – are incorporated into courses;
- the extent to which modelling has been the focus of professional development;
- the extent to which modelling activities have been incorporated into textbooks and other curriculum resources.

3 CURRENT SITUATIONS

The author conducted a comparative study in 1992, seeking information from each state on these matters. A personal evaluation of important features of the responses is as follows.

Mathematical modelling in New South Wales

Assessment in the most populous state is still completely examination dominated, despite an expressed intention to consider introducing mathematical project requirements into some post-1994 courses. The following extract from the Mathematics Curriculum Committee's response (private communication, 1992)

> most courses, except those of the highest levels, should be presented more concretely and less formally than at present

suggests that mathematical application, let alone modelling, is still not seen as a key objective for courses for the more gifted students.

Mathematical modelling in Queensland

All the new senior secondary mathematics courses currently being introduced in Queensland have a strong modelling focus. In one mainstream course, for example students should be able to

- understand that a mathematical model is any mathematical representation of a situation
- identify the assumptions and variables of a simple mathematical model of a situation
- form a mathematical model of a life-related situation, over a variety of areas of application
- derive results from consideration of the mathematical model chosen for the particular situation
- interpret results from the mathematical model in terms of the given situation
- explore the strengths and limitations of a mathematical model
- develop a set of procedures to be used in tackling modelling problems

which is an extract from the revised draft of the Senior Syllabus in Mathematics A, Board of Senior Secondary School Studies, Queensland (1992).

Strong leadership in implementing these objectives is provided by the continuing work of Clatworthy and Galbraith (1991), who continue to provide teachers with exemplary modelling tasks and a curriculum and assessment framework which is being widely adopted, the more so with directed investigations and project work being built into the new course requirements.

Assessment criteria for modelling projects

Ability to specify problem clearly	*Ability to set up a model*	*Ability to solve maths, interpret, validate and refine*	*Ability to communicate results*
Standard 1 Is able to proceed only when clues are given	Is able to proceed only when clues are given	Is able to solve the maths problem when given clues and hints	Can communicate reasonably - in presentation - in conciseness - in oral form
Standard 2 Can extract clues from information and clearly define the problem	Is able to develop the model with minimum assistance	Is able to solve the basic problem with no assistance	Can communicate clearly
Standard 3 Can extract clues and clearly define the problem when information is open-ended, insufficient or redundant	Is able to develop the model independently where no clues exist	Can independently solve the problem and can evaluate and refine the model	Can communicate clearly with innovative and creative ideas

Figure 1. Adapted from Clatworthy (1989)

Queensland has a long tradition of internal standards-referenced assessment within state-wide supervisory structures, allowing performance on modelling objectives, as distinct from familiarity with models, to be assessed over a range of activities. Within a state-wide framework, with appropriate checks and balances, a randomly selected peer checks a school's intended curriculum and proposed assessment tasks. Later in the year this person works again with the school in assessing student reports and allocating grades. Generally teachers respond positively to this local autonomy and professional responsibility although, with fewer quality controls than other state's assessment arrangements, some of the assessment tasks used (such as Tiling your Bathroom) have very little modelling requirement.

Mathematical modelling in Western Australia and Tasmania

Modelling objectives feature strongly in both these states, but at this stage only in those lower-level courses for which assessment by means other than examinations has been accepted. In Western Australia the new Modelling with Mathematics Course (Ministry of Education, 1991) has, in its first year, attracted approximately 30% of final-year students. Modelling objectives are weighted 60% in overall assessment, which must include 50% to 70% of project work. The state education authority provides detailed support material for the course, and a content framework from which teachers choose particular classroom investigations. For each investigation the supporting documentation includes not only such things as focus questions, suggested report headings and appropriate secondary data, but also cross-referenced lists of appropriate modelling objectives, content objectives specific to the topic, and general numeracy objectives. Teachers work confidently within this framework, giving due weight to each of these different sets of objectives.

Mathematical modelling in South Australia and the Northern Territory

Mathematical modelling in Australian senior secondary classrooms started in the small state of South Australia in the early 1970s. Since that time a range of high-quality modelling activities can be found in South Australian and Northern Territory classrooms, supported by documentation from the assessment authority (SSABSA 1991, 1992) and the professional associations (see Finger and Treilibs, 1992).

The following is an extract from the conclusions section of a student report of an applied geometry project.

We believe that we have successfully found a direct line of sight for the microwave link, namely between the wheat silos at Kingscote and the sheds in Parndana. However for this to be an unbroken microwave link we would have to build a tower on top of the sheds.

Belinda calculated that by assuming the Earth was flat there was not a direct line of sight so she worked out how high a tower, best placed at Parndana, needed to be built.

Denise showed that using the Earth's curvature and the correct topographical features she could not get a direct line of sight, so she then calculated how high a tower they would need to build on top of the sheds at Parndana. This answer came to 50 metres, so that the observer from Kingscote would have to look up at an angle of 0.003 degrees.

There was a difference of 11 meter between Belinda's and Denise's results for the height of the tower, showing the effect of the curvature of the Earth.

Following initiatives from Victoria, the new South Australian Certificate of Education promises to provide stronger support for assessment of extended tasks, with extension of modelling objectives into a wider range of available mathematics courses. Directed investigations, project work and mathematical modelling are identified as major features of all penultimate-year courses, and the Assessment Board employs a team of moderators who liaise with schools throughout the year to

- develop an agreed memorandum of assessment principles and practices,
- endorse the assessment plans proposed by the school,
- discuss and advise on implementation of assessment arrangements,
- look at samples of student work to check teachers' interpretation of assessment standards.

This tradition of close liaison between Board personnel and classroom teachers helps ensure that experience is passed on effectively, and a good standard of curriculum implementation is maintained. Most of the schools are reasonably centrally situated and the programme is regarded as cost effective. Extension of similar arrangements into the new final-year courses is, as in Western Australia, strong in some courses – notably Applied Mathematics – but not in others. Courses involving competition for tertiary entrance involve statistical moderation of non-examined work, with less-tightly specified extended-task requirements and less incentive for modelling activities.

Mathematical modelling in Victoria

The down-side of modelling in all of the above situations is that courses for the more able students have not as yet been changed along similar lines. While there is general support for these changes, the development of appropriate assessment for the more competitive courses remains a stumbling block. A breakthrough has been made in Victoria, where problem-solving and modelling activities constitute one of three major work requirements in every mathematics course within the Victorian Certificate of Education (VCAB, 1990, 1992; Money and Stephens, 1993). Centrally-set assessment tasks involve all final-year mathematics students, two-thirds of the age cohort, in extended investigation (up to four weeks) of project topics and challenging problems. Although limited in scope, due to accessibility and resource requirements, these include significant opportunity for modelling activity, even though it is not clearly differentiated from more pure-mathematical problem solving.

The following is taken from a two week common assessment task for a Victorian calculus course, Victorian Curriculum and Assessment Board, 1991.

> *Walking in the rain*
>
> On a rainy day a person with an umbrella is walking North at a speed v m/s. A wind is blowing from the South and rain is falling with speed 2 m/s at an angle α to the vertical.
>
> At what angle β to the vertical should the person position the umbrella so that it blocks out the maximum amount of rain? Under what conditions will the umbrella need to be held vertically upwards?
>
> Unfortunately this person has to walk for a long time in the rain. Eventually the umbrella becomes saturated and rain begins to pass through it. However it is possible that by varying the speed of walking, the amount of rain falling on the umbrella can be kept to a minimum.
>
> If the umbrella is positioned at angle β, as calculated previously, what is the desired walking speed which will minimise the amount of water falling on the umbrella?

The significant contribution of these assessment tasks in highly-competitive tertiary selection has ensured the production of some excellent-quality student work, new-style student texts, and major improvements – despite the lack of adequate support – in teachers' classroom practice. Introduction of the 'common assessment task' approach, common to all subjects, has not been without difficulties, some of which are specific to mathematics. A major problem is a divided tertiary mathematics-education community where, with some notable exceptions, the educators support the changes and the mathematicians continue to emphasise content-skill requirements and exclude modelling activities from their own first-year courses.

4 CONCLUSIONS

The Victorian experience most clearly illustrates the adage "What you test is what you get". With this in mind, different Australian states have made strong efforts to incorporate assessment modes which best promote modelling objectives. The fact that modelling is less often incorporated in the more competitive courses points to a need for further research towards development of a profile of progress in the learning of mathematical modelling skills. This research, together with promotion of modelling in undergraduate mathematics-related courses, is desperately needed to gain support for modelling objectives from the tertiary education community. The current international debate, such as through the 'Moving Beyond Myths' initiatives in the USA, is of vital assistance. The Western Australian work provides a major contribution in demonstrating, through its detailed documentation, that both process and content objectives can be met through the same learning activity.

Turning modelling objectives into implemented curriculum requires continued support from educational systems, from the professional associations, and from tertiary educators. The benefits of many years of such cooperation in Queensland and South Australia are now being unleashed by the structural changes in certificate requirements. The Modelling with Mathematics initiatives in Western Australia are an example of what Geoffrey Howson called for at ICME-7, when he said "Give teachers a framework within which they can demonstrate their professionalism".

REFERENCES

Australian Education Council (1990). *A National Statement on Mathematics for Australian Schools*. Melbourne: Curriculum Corporation.

Clatworthy, N (1989). Assessment at the upper secondary level. In Blum, W *et al* (ed), *Applications and Modelling in Learning and Teachiong Mathematics*. Chichester: Ellis

Horwood, 60-65.

Clatworthy, N and Galbraith, P (1991). Mathematical modelling in senior secondary mathematics: implementing an innovation. In *Teaching Mathematics and its Applications,* **10**, 6-22.

Finger, J and Treilibs, V (1992). *Applied Mathematics Problems: a Teacher Resource.* Adelaide: Mathematical Association of South Australia.

Mayer Committee (1992). *Employment-Related Key Competencies: a Proposal for Consultation.* Melbourne: Mayer Committee.

Ministry of Education, Western Australia (1991). *Modelling With Mathematics.* Perth: Ministry of Education, Western Australia.

Money, R and Stephens, M (1993). Linking applications, modelling and assessment. In de Lange, J *et al* (ed), *Innovation in Maths Education by Modelling and Applications.* Chichester: Ellis Horwood, 323-336.

Money, R and Stephens, M (1993). Meaningful grades for assessing extended tasks. In Webb, N (ed) (in press), *Assessment in the Mathematics Classroom. 1993 Yearbook of the National Council of Teachers of Mathematics.* Reston: National Council of Teachers of Mathematics.

National Research Council (1991). *Moving Beyond Myths: Revitalising Undergraduate Mathematics.* Washington: National Academy Press.

Senior Secondary Assessment Board of South Australia (1991). *Mathematics Stage 1 Extended Subject Framework.* (1992) *Mathematics Stage 2 Extended Subject Framework.* Adelaide: Senior Secondary Assessment Board of South Australia.

Victorian Curriculum and Assessment Board (1990). *VCE Mathematics Study Design.* Melbourne: Victorian Curriculum and Assessment Board.

Victorian Curriculum and Assessment Board (1992). *VCE Administrative Handbook 1993.* Melbourne: Victorian Curriculum and Assessment Board.

16

Mathematical modelling in an extended engineering course

Edward Graham
University of Plymouth, UK

SUMMARY

This paper describes two examples of mathematical-modelling projects undertaken by first-year engineering students using the DERIVE computer algebra package. The examples show how students can use the graphics facility to gain insights into the weaknesses or behaviour of models, how derivatives and limits can be found with accuracy and confidence, so that the students can concentrate on the modelling problem and not become lost in the mathematics. The result of incorporating the use of this package is greater student confidence, and a better realisation of their ability to apply mathematics.

1 INTRODUCTION

This paper will concern itself with the mathematical-modelling component of the extended engineering course run at the University of Plymouth, and show how the use of the symbolic algebra package DERIVE (Soft Warehouse, 1992) plays an important rôle in the modelling work undertaken by the students on this course.

The primary aim of the extended engineering course is to prepare students so that, after one year of study, they can commence one of the University's engineering degree or diploma courses. Mathematics forms a substantial component of this course, alongside physics and a small amount of engineering experience. The mathematics is taught in a number of small units, each of which is individually assessed, plus a mathematical-modelling project that is carried out in the final term of the course.

Most of the students who take this course have very weak mathematical backgrounds, and have not studied mathematics since the age of sixteen – a long time for the many mature students on the course. In order to meet the needs of these students, the mathematics teaching has been designed to make the content of the course accessible to these students. One significant way in which this has been achieved is through the extensive use of the computer algebra software DERIVE. The content of the course is not unusual, including work on algebraic manipulation, differentiation, integration, trigonometric functions and series, plus applications of these areas of mathematics, particularly in mechanics. The course contains a number of prescribed activities that are carried out with DERIVE, for example an introduction to differentiation or combinations of trigonometric functions, which can be found in Watkins (1992). However, since the computing facilities are very accessible, it is also possible to use DERIVE in a more spontaneous manner, when it can assist greatly in the students' learning. The students are also encouraged to make use of the DERIVE facility in their own time, when tackling assignments and other tasks. An essential feature of the use of DERIVE in this course is that it is used to provide an investigative environment, rather than as a tool that simply produces answers. Thus the expectations of the students are that DERIVE will help them to explore mathematics, and improve their understanding, as well as providing answers to specific problems. This is reflected in the way that they use the package in their own time.

One of the mathematics units is a short course in mechanics. This is used as a vehicle to provide an introduction to modelling. As ideas are introduced and developed they are treated as mathematical models, and substantial emphasis is placed on the formulation stages of problem solving. Also, as part of this course, a problem-solving procedure is developed, which contains many of the stages of the mathematical-modelling cycle. This work on mechanics helps introduce the students to some of the elements of mathematical modelling before its timetabled slot in the final term, and many students ultimately choose modelling projects from this area.

The bulk of the final term is devoted to project work. The students carry out an engineering project in the discipline of their choice, as well as the mathematical-modelling project. The students are introduced to the fundamental aspects of mathematical modelling during the first few weeks of term. During this time they work on a number of small tasks, initially of an empirical nature and then moving on to deal with theoretical models. While working on these tasks the students are encouraged to make use of DERIVE as and when it is appropriate. The students then spend a longer period of time working on a problem that they choose from selection on offer, or they can submit their own problem. This project is assessed by a written report and an oral presentation.

This paper will now consider examples of projects carried out by these students, and show how students have made use of DERIVE to assist their modelling. Specific examples of this will become apparent as the students work is examined, but some general comments can be made first. The students have progressed from a very weak starting point, learning a great deal of mathematics in a very short time, and are now applying this mathematics in a modelling context, often selecting the appropriate mathematics themselves for the first time outside a more formal teaching situation. They use mathematics that is new and still fresh, but not deeply consolidated, suffering from what is sometimes termed the 'one year gap', between the time when they learn a piece of mathematics and the time when they are confident to use it in a problem-solving situation.

2 PROBLEM 1: *MOTORWAY TRAFFIC JAMS*

This is a version of a well-known problem that can be found in a number of problem banks, for example that of the MEW Group (1989).

Problem statement

Traffic queues often form on motorways when three lanes of traffic are forced to reduce to two or one lane because of road-works. This is a major frustration to motorists, and may extend journey times by hours. Consider the problem of reducing two lanes of traffic to one. What speed of traffic achieves a maximum flow through the motorway section?

Student approaches to the problem

Students usually begin by assuming that all the cars travel at a constant speed, and are all the same distance apart. If a vehicle enters the motorway section every t seconds, then the number of cars entering the section per second is $\frac{1}{t}$. As the vehicles travels at a constant speed, v, the distance, d, between vehicles is given by $d = vt$.

From this expression we have

$$\frac{1}{t} = \frac{v}{d}$$

Then the flow rate (in cars per hour) is modelled by

$$F = 3600 \cdot \frac{1}{t}$$
$$= \frac{3600v}{d}$$

where d is the distance between the fronts of two vehicles. In fact d is made up of two parts – the length of the vehicles (all assumed to be the same) and the actual separation between them. In the UK the government publishes recommended separation distances for driving at various speeds in good driving conditions. As part of the empirical modelling work, the students have already demonstrated that these distances can be modelled by

$$s = 0.682v + 0.076v^2$$

A common approach is to assume that the length of the vehicles is zero, as this is small compared to the separation distance, to give a flow rate modelled by

$$F = \frac{3600v}{0.682v + 0.076v^2}$$

Some students find it difficult to visualise how this flow rate changes with v – particularly because, if they substitute values for v, they find that F decreases as v increases and that F increases as v decreases, with a flow rate of over 5000 cars per hour when $v = 0$! I have seen students struggle to come to grips with this, often reaching the conclusion that their calculations are wrong, and seeking assistance with the substitution of values and calculation of F, or question their model for the distance between cars. However if the students move to DERIVE they can quickly plot F and have confidence that what they produce is accurate. On the screen, as shown in Fig. 1, they can see their expression for F, the graph that the package produces, and can calculate F for any v to confirm their original predictions. In this case DERIVE has helped the students to move more rapidly towards the conclusion that their model is in need of revision.

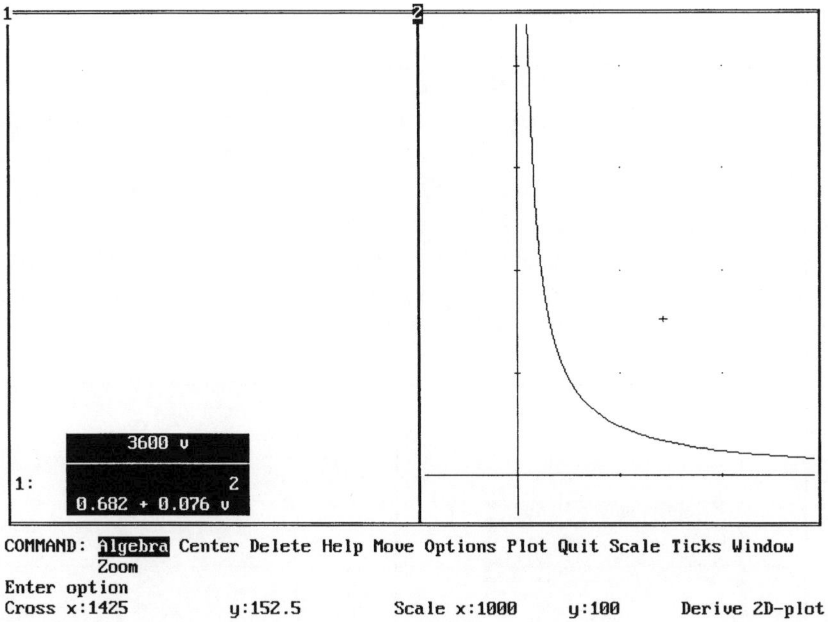

Figure 1

Once the students have got to this stage they can re-examine their assumptions, and they often decide to include the length of the vehicle. A typical modification would be that they include the length of a car as 4.5 m, to give a new model for the flow rate of

$$F = \frac{3600v}{4.5 + 0.682v + 0.076v^2}$$

With DERIVE, a fresh plot shows a curve that does give a zero flow rate at zero speed, and appears to have a maximum for some value of v. I have seen students without DERIVE laboriously produce a hand-drawn graph and even then be uncertain about its accuracy. With the curve on the screen in front of them, the students may position the cursor on the maximum and read off the maximum flow rate and the corresponding speed. Alternatively they will try to use calculus to find these values. The differentiation of the flow rate requires an application of the quotient rule which is not really difficult, but for students who have only recently learned this rule an expression like this can be quite daunting. DERIVE can produce quickly the derivative for them, and then find when this is zero. Figure 2 shows the plot, and the differentiation. This process would not trouble a mathematician, but can be very time-consuming and confidence draining for students who are relatively new to mathematics at this level. I have had students bring me several possible versions of the derivative, and ask me to identify the correct one. DERIVE allows the students to differentiate with confidence, applying the concepts that they understand. Most students would also want to prove that they can also obtain the correct derivative by hand. Clearly the approach described above is not the only way to solve this problem – an alternative approach would be to minimise the time interval between cars. However, as DERIVE has the ability to handle a wide range of mathematics, it would clearly be possible to utilise the package in different ways with different approaches.

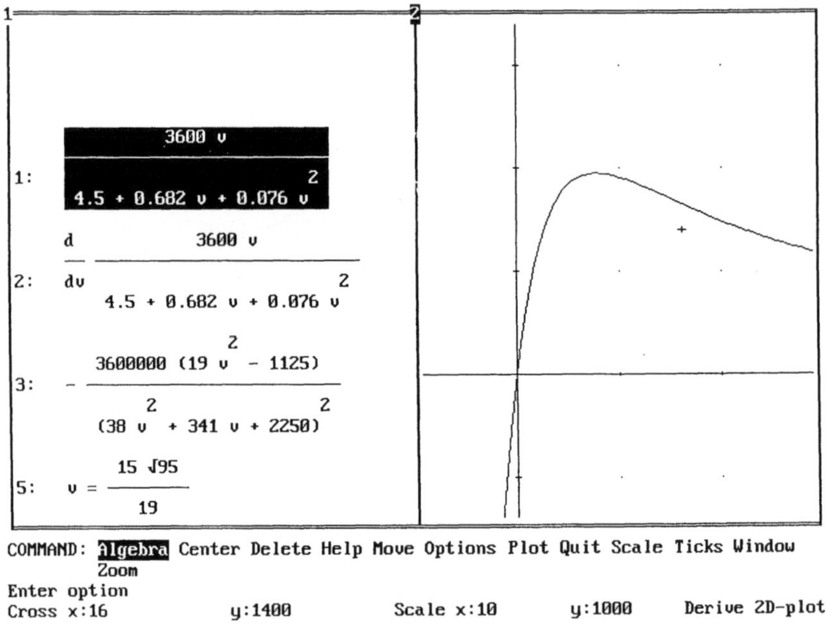

Figure 2

This problem illustrates how DERIVE can build students confidence and help them to concentrate on the modelling process, showing them that they have the ability to use mathematics to solve this problem. Not only does this help their confidence, but it enhances their enjoyment of their project work, making it more likely that they will want to use mathematics to solve other problems that they encounter in their studies.

3 PROBLEM 2: *HOW LONG ON GREEN*?

This is a problem that was generated by a student, after he had attempted other traffic-flow problems in an earlier part of the course. At the outset he expected to form a model that would lead to an expression that he could differentiate to obtain an optimum value for the timing of traffic lights, similar to that obtained for the previous example.

Problem statement

A common situation on major roads is that one side of the road is closed to traffic owing to road works. What sequence of timings should the traffic lights use to ensure the maximum flow rate?

Student approaches to the problem

The students working on this project produced a first model, and then used DERIVE, to enable them to understand their model and make predictions using it. Their basic assumptions were that all cars waiting to pass through the traffic lights occupied 5 m of road space, they would accelerate at 1 ms^{-2}, and that cars would begin to move $\frac{2}{3}$ s after the car in front. They defined the flow rate as

$$F = \frac{n}{t_n}$$

where n is the number of cars and t_n is the time taken for n cars to pass through the lights. They produced a model for t_n based on their assumptions, made up of two parts – the total of the drivers' reaction times and the time taken for the last car to travel through the lights. The reaction-time component was simply formed by assuming that each driver took $\frac{2}{3}$ of a second to react, and so for n cars the total reaction time would be $\frac{2n}{3}$. The time taken to reach the lights was based on a common acceleration of $1ms^{-2}$, and the assumption that the first car had to move 5 m to clear the lights, the second 10 m, the third 15 m, and so on. The distance travelled by the nth car is then $5n$ metres. The well-known constant-acceleration formula

$$s = ut + \frac{1}{2}at^2$$

was then applied. As $s = 5n$, $u = 0$ and $a = 1$, the time to travel past the lights was obtained by rearranging

$$5n = \frac{1}{2}t^2$$

to give

$$t = \sqrt{10n}$$

Combining these two components gives

$$t_n = \frac{2n}{3} + \sqrt{10n}$$

and a flow rate of

$$F = \frac{n}{\frac{2n}{3} + \sqrt{10n}}$$

The students had produced an expression that they wanted to differentiate, but which was not straightforward. It would certainly have put their skills of algebraic manipulation to the test, but would have done little to move their modelling project forward. Their natural resort was to turn to DERIVE for some assistance, where they were quickly able to produce the derivative and plot it as shown in Fig. 3.

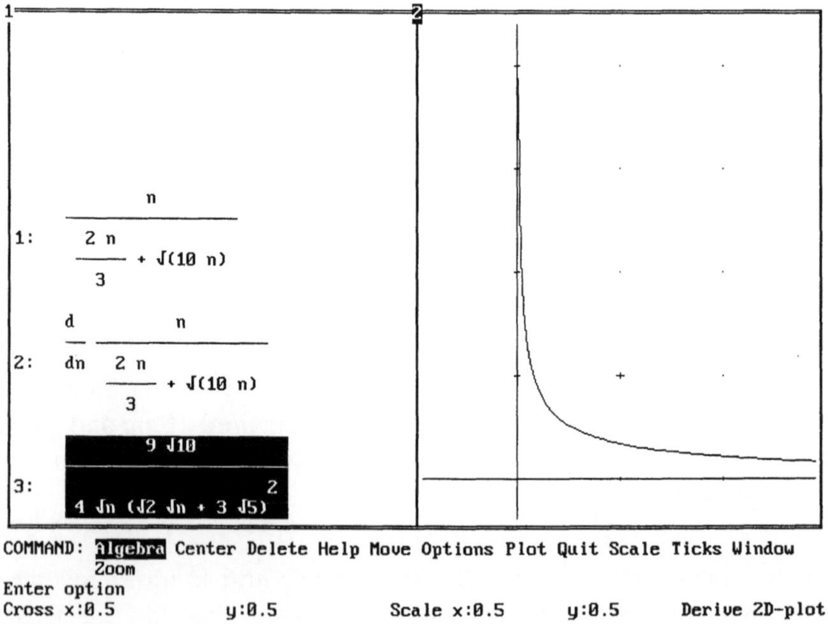

Figure 3

They were able to see that the flow rate did not have a local maximum, so they decided to produce a plot of the function. If they had done this by hand they would have been faced with the question of what happens when $n = 0$; this too could be a concern of students working with DERIVE. The students on the course had done work on limits, and were able to ask DERIVE to find the limit as n tends to 0. This is shown in Fig. 4, along with the plot of the function. When the students saw the graph they were able to visualise what was happening as n increased, and began to question their initial assumptions. The students tried various approaches to improve the model. One was to assume that the cars never exceeded the speed limit for a built up area, 30 miles/h (or 13 m/s). A further consideration was to include the effects of a change-over time, as traffic from other roads passed through the junction. The students did, however, check out their original model by monitoring the flow of cars through a busy junction, and observed that the model gave good predictions for the flow from one direction.

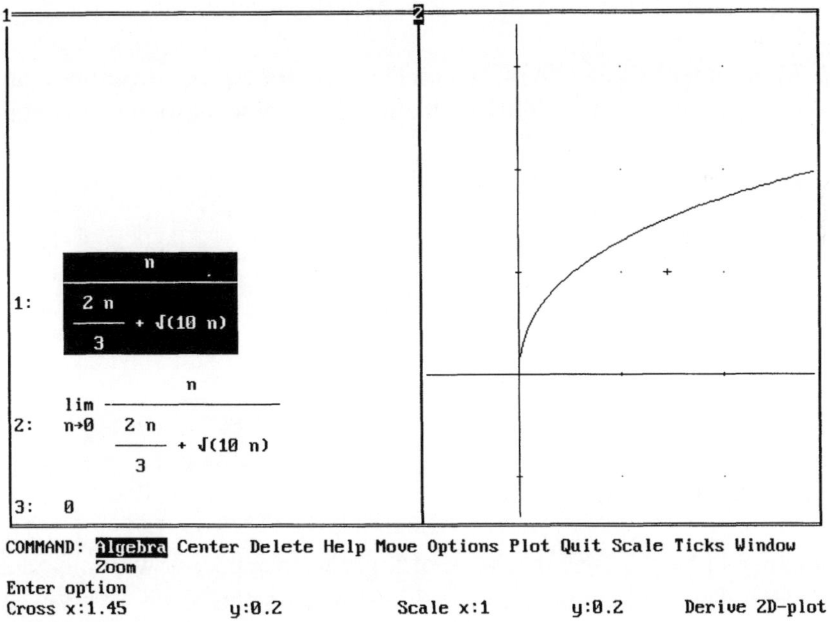

Figure 4

Here is another example where DERIVE has helped students to understand the mathematics that they are dealing with. The students working on this project were disappointed to find that they could not maximise the function that they had produced in their model but, with the help of DERIVE, they realised that they should question their assumptions and not their mathematics.

4 THE BENEFITS OF *DERIVE* FOR STUDENTS

I have observed four main ways in which packages such as DERIVE can benefit students.

- Computer-algebra systems have allowed students to use concepts that they understand, but find difficult to apply. In the two examples discussed above, they have used differentiation and taken limits, which they would have found very difficult unaided.

- Computer-algebra systems allow students to explore solutions to problems graphically. This can help students to understand the weaknesses of the behaviour of their models, as seen in both examples.

- More emphasis can be placed on the modelling process, rather than on the solution of a mathematical problem. Often students spend a disproportionate amount of time finding solutions, rather than modelling, and this can be resolved with a package such as DERIVE.

- Many teachers refer to the 'one year gap', corresponding to the time between when students meet a concept and can use it effectively in problem-solving. This gap can be reduced by the use of a suitable computer-algebra system.

5 CONCLUSION

This paper has discussed the work of a group of students who have learned a great deal of mathematics in a very short time, starting from a very weak base. A short mathematical-modelling course provides them with a valuable problem-solving experience that can utilise the mathematics that they have been learning. However, their confidence is still at a stage where it needs to be nurtured, and so the modelling work has to be handled with care. Using DERIVE takes away much of the risk of failure, providing an environment in which they are familiar and can explore the mathematical equations that they produce with confidence.

The result is that the students see themselves doing useful mathematics, their confidence grows, they feel comfortable, and are enthusiastic to do mathematics. One regular outcome is that some students opt to study mathematics at the end of the course, rather than engineering, because their experience of mathematics – in particular during the modelling component – has been so positive.

This paper has described the use of DERIVE at a fairly low level, that does little to draw upon the power of the system, but it has done a great deal to boost the confidence and morale of these students. Clearly, as students learn more sophisticated mathematics, and use it in modelling situations, DERIVE could play a similar rôle, but at a higher level.

REFERENCES

MEW Group (1989). *Exploring Mechanics.* London: Hodder and Stoughton.

Soft Warehouse (1992). *DERIVE, A Mathematical Assistant.* Honolulu.

Watkins, AJP (1992). *DERIVE Based Investigations for Post-16 Mathematics.* Bromley: Chartwell-Bratt.

17

Industrial applications of mathematics for engineering students

Heinrich Abel
Deutscher Institut für Fernstudien, University of Tübingen, Germany

SUMMARY

This paper reports on a further-education programme in mechanical and manufacturing engineering for practising engineers. Three courses deal with the basic elements of feedback control technology, applied position control technology, and computer applications in manufacturing. In this context, mathematical modelling of dynamical systems plays a prominent rôle.

The programme was devised as a bilingual multimedia course, to be supplemented by face-to-face teaching. It consists of videotaped lectures as the main medium, and accompanying written material which serves as an introduction to the video lecture and expands on important aspects. Like the lecture tapes, the written material is bilingual (English and German).

1 PRESENT SITUATION

The state government of Baden-Württemberg (Germany) regards international cooperation in the field of science and technology as a matter of great importance, particularly to cope with the technological and economic challenge of the future. One example is the government's policy to support and to promote high-technology cooperative projects between higher education and research institutions in Baden-Württemberg and similar institutions in highlyindustrialised European and non-European countries. Computer applications in mechanical and manufacturing engineering is one of the fields where this challenge is particularly evident.

There is a strongly-felt need for upskilling practising engineers, especially those who have been in the profession for a long time and in fields which are characterised by a rapid development of science and technology. World-wide, modern data-processing techniques are now intensively used in mechanical and manufacturing engineering. Feedback control technology using computers is of particular significance here. Various surveys have underlined the need for continuing advanced level education of practising engineers in industry, covering electronics, computer science and a range of applications in all spheres of mechanical engineering. Companies working in this field have indicated that they would welcome courses by universities and similar institutions, provided these are flexible enough not to require a prolonged absence of employees from their workplaces. Emphasis should be on individualized forms of learning, combining self-study using various media with phases of face-to-face instruction.

In recognition of these needs, and in the framework of international cooperation between the states of Baden-Württemberg (Germany) and Massachusetts (USA) it has been agreed that institutions in both states develop jointly a bilingual training programme for practicing engineers on computer applications in mechanical and manufacturing engineering, based on videotaped lectures.

The institutions who cooperate in this project are the following.

- German Institute for Distance Education at the University of Tübingen (DIFF)
  Project management, conceptual work, advising authors of the videotaped lectures and written material, editorial treatment of manuscripts to facilitate the use of the texts as self-study material

- Institute for Control Technology, Fachhochschule für Technik Esslingen (FHTE)
  Course I: Basic elements of feedback control technology

- Institute for Control Technology of Machine Tools and Manufacturing Systems (ISW), University of Stuttgart
  Course II: Applied position control technology

- Institute for Real-Time Computer Systems and Robotic (IRP), University of Karlsruhe
  Course III: Computer applications in manufacturing

- Technical Academy Esslingen (TAE)
  Production of videotapes for Course II, project management, distribution of the material in Europe, seminars with face-to-face instruction

- Video Instructional Program (VIP), University of Massachusetts at Amherst
  Production of videotapes for Course I and III, distribution of the material in North America

2 OBJECTIVES OF THE PROJECT, CONTENT, AND TARGET COMMUNITY

Objectives

The project has three major objectives.

- To provide practicing engineers who acquired their university or college degrees a long time ago with those qualifications and skills which they need in order to use computer-based manufacturing systems, and to adapt and develop their skills in accordance with varying production needs.

- To improve the German engineers' technical English and the American engineers' technical German – not an unimportant aspect bearing in mind the extent of the international transfer of know-how and technology – by presenting bilingual material. (The German part could also be replaced by another language. There are already demands from South America for a Spanish and a Portuguese version.)

- To provide continuing education through an (experimental) form of teaching and learning which combines self-study using various media with phases of face-to-face instruction, thus permitting a high degree of individualised learning. This should be compatible with the living and working conditions of professional adult learners and the needs of the companies where they work.

Contents

To serve these objectives, material is being developed for three courses.

Course I: Basic Elements of Feedback Control Technology

Module 1:	Introduction to continuous-time control
Module 2:	Descriptions of control loop components
Module 3:	Modelling of transfer systems
Module 4:	Connection of the plant and the controller within a control loop
Module 5:	Controller design and control loop synthesis
Module 6:	Introduction to linear, discrete-time systems
Module 7:	Modelling of discrete-time control loops
Module 8:	Performance of closed-loop discrete-time control system
Module 9:	State variable models for linear time-invariant systems
Module 10:	Control synthesis in state space

Course II: Applied Position Control Technology

Module 1:	Basic elements of position control and mathematical models of the plant
Module 2:	Design of mechanical transfer elements
Module 3:	Structure of electrical feed drives
Module 4:	Parameter optimisation and path behaviour of classical control loops
Module 5:	Modern methods of position control
Module 6:	Computer-aided design of control systems

Course III: Computer applications in manufacturing

Module 1:	The manufacturing systems
Module 2:	CIM models and concepts
Module 3:	Analysis tools for manufacturing
Module 4:	Flexible manufacturing and assembly systems
Module 5:	Control architectures of manufacturing systems
Module 6:	Communication nets and protocols
Module 7:	CAD: its rôle in manufacturing
Module 8:	Process planning and manufacturing scheduling

Module 9: Diagnosis of manufacturing systems
Module 10: Robotics
Module 11: Material handling
Module 12: Quality assurance

Target community

The material of courses I - III will be used primarily in different continuing education offerings. The target group are engineers with university qualifications, engineering practice, or technical school degrees in the areas of manufacturing engineering, electrical engineering, electronics and data processing, as well as engineers who wish to become familiar with new technology and engineers who wish to review their knowledge. In addition, the courses are directed generally at electrical and mechanical engineering students.

3 THE MULTIMEDIA PACKAGE

The contents of these courses are presented in a multimedia package.

The main medium is the *video lecture*, delivered by an expert of the respective field at a university or college. The lectures provide a core, which is -where necessary- augmented with computer simulations, laboratory experiments or industrial demonstrations. The videos are taped with two-track sound, – one track contains the German sound, the other contains the English version. The learning effectiveness in working with the video is at its best when the material presented is adjusted to the individual's own learning pace. For this purpose, the tape may be stopped and replayed if necessary in case the subject has not been understood during the first viewing. In addition, the study commentary should be used to accompany the viewing.

An *Introductory Text* and a *Study Commentary* help prepare for the video lectures and deepen the knowledge obtained there. In the study commentary the most important contents of the video lecture are repeated, and some additional or more in-depth information is given. In addition, the study commentary provides longer mathematical derivations, formulae, tables, and figures which may be referred to in the video lecture. Owing to this close interaction between the video lecture and the study commentary, the study commentary is not a systematic, comprehensive and coherent textbook. Each Study Unit ends with some review questions. In addition to these review questions, the courses also contain a collection of problems. Their purpose is to evaluate the success of learning. The 'problems' have been chosen from a wide range of industrial subject areas. The study commentary also contains references to relevant literature.

Apart from independent video-supported study phases for individual learners, *group meetings* should also be provided at certain intervals for students to benefit most effectively from the multimedia package. During such seminars, the engineers will get to know hardware and software, exchange their experiences, and discuss questions to which they did not find answers during their self-study.

4 ENGINEERING TOPICS, RELEVANT MATHEMATICS, AND MATHEMATICAL MODELLING

Engineering topics

As an example, the contents of Course I will be listed in detail. This course is composed of the Modules described below, which may be divided into three gropus.

Modules 1 to 5: Analogue feedback control
Modules 6 to 10: Digital feedback control
Modules 9 and 10: Feedback control using state variable models

Module 1 introduces continuous-time control. The task of process control is given in a general form, and the basic construction of a feedback system, with the control equipment, plant and measurement equipment, is explained.

In the following modules the individual topics mentioned in Module 1 are treated in more detail, and the necessary mathematical analysis methods are discussed.

In *Module* 2 the description of individual transfer elements in process control is demonstrated. The result of this description is the so-called *block diagram* and the *transfer function* of a system.

The topic of *Module 3* is *modelling*. By modelling – also called model construction – we mean the systematical analysis of a transfer system using block diagrams. It is enlightening that it is not possible to design a controller – whether of analogue or digital design – until the plant component to be controlled is exactly known and described.

Module 4 describes the interaction of a plant - the component which is to be controlled – with the time-continuous control equipment. Special emphasis is placed on the stability of feedback systems.

The topic of *Module 5* is controller design. It deals with the question of how to choose the correct controller for a particular component to be controlled, and how to design its parameters. In general it is not sufficient simply to guarantee stable behaviour. For example, a good controller design carries out the compensation of a disturbance without overly-large overshoot.

In *Module 6* discrete-time process control is introduced. After the description of digital signals, the *Z-transform* and its application to discrete feedback systems is the central topic of this Module.

Module 7 deals with the modeling of discrete-time feedback systems. Here the most important basic elements of process control are derived in their discrete-time form. Special attention is given to the analysis and description of the sampling process, and the conversion of analogue signals to digital signals.

In *Module 8* (corresponding to Module 4), the interaction of digital control equipment with a plant is discussed. The influence of the *sampling time* of the controller, which influences the stability of the overall feedback system to a large extent, is covered. The sampling time is also of decisive importance in the design of a digital controller.

The *transfer function* or the *frequency domain* were the basis for the analysis of single-loop feedback systems in modules 1 to 8. However, if multiple-loop feedback systems – feedback systems in which multiple input and output signals occur – are treated, the technique of state description of systems is required. This type of system description is the subject of *Module 9*. The examples demonstrate that the technique of *analysis in state space* can also be used to advantage in single-loop feedback systems.

In the concluding *Module 10*, the possibility of systematic controller design concepts is demonstrated, based on a state description. The *state controllers* developed from this do not have the simplicity of a PI- or PID-controller. However, realisation of such state controllers with microcomputers is no longer a problem today.

Relevant mathematics

In the first 8 modules the mathematics of dynamical systems, where differential and difference equations occur, plays an important rôle. Modules 9 and 10 require a profound knowledge of linear algebra – matrices, eigenvalues, bilinear forms and so on. These concepts are explained step by step where they are needed. Thus engineers who are not familiar with the mathematics involved should be able to understand the relevant topics.

Mathematical modelling

As stated above, modelling in this context is the analysis of a dynamical system (transfer system) using block diagrams. If necessary, such a block diagram, which at first is only designed on paper, can also be transformed into an actual model. Examples of this are the construction of a small scale automobile model, or the calculation of the system response of a block diagram using analogue or digital computers.

The key to the solution of process-control problems is modelling the transfer systems. Implicit with modelling is the limitation of a transfer system to that which is significant. This often permits a reasonable first analysis, and from that the directed design of control systems.

It is often necessary to linearise nonlinear transfer elements at a suitable operating point. Furthermore the block diagram will usually be normalised. This means elimination of dimensions, thus yielding more easily understood block diagrams. After linearisation and normalisation a model of the dynamical system is obtained, from which the differential or difference equations that describe the system mathematically can be derived.

To handle those equations one can apply several methods.

- The equations could be solved exactly, using the methods of Laplace transform for the time-continuous case or Z-transform for the discrete-time case. By such a transformation from the time range to the Laplace or the z-range a transfer function $F(s)$ or $F(z)$ is obtained, which describes the behaviour of the control loop. $F(s)$ or $F(z)$ are rational functions (assumed that delay time is not taken into account) whose zeros and poles can easily be calculated. The location of zeros and poles in the complex plane provides information on the behaviour of the control loop – such things as stability, resonance, damping.

- Another way analysing a control loop would be simulating it's behaviour on a computer. As powerful simulation programs are nowadays available even for PCs, this would be an easier and quicker way.

- One can also solve systems of linear differential equations by Laplace transforms via the Leverrier-Fadeev algorithm. This method can be used to give a very simple and general treatment of linear systems with an absolute minimum of theory from linear algebra, provided one has access to a computer-algebra system to perform the necessary calculations.

Reflecting on the mathematical methods mentioned above leads to the general question of how much mathematics engineers really need at their work place. There are different views on this, even among the authors of the course. The evaluation of the course material, which will be carried out in 1993/94 may lead to a more reliable answer.

An example of mathematical modelling: the suspended ball

The following example of a metallic body suspended under a solenoid plays an important rôle in various technical applications; the underlying principle is applied, whereever an electromagnetic-mechanical converter is employed (e.g. relays, electrovalves). One of the most conspicuous technical applications of this principle is a magnetic levitation vehicle.

This principle is treated in Course I, using different approaches; first in Module 3 (using classical methods of feedback control technology; even in the 1930s this was still regarded as impossible) – and then again in Modules 9 an 10 (using modern methods of state variable models).

We will consider a system which consists of a metallic ball suspended underneath an electromagnet (see Fig. 1). The external inputs to this system are the voltage across the coil of the magnet (which can be manipulated), and the acceleration of the ball due to gravity. Two outputs are available – the voltage from a light-sensing diode, which is assumed to be proportional to the height of the ball underneath the magnet (measured relative to the steady state position of the ball) and the current through the coil. The magnet is assumed to have inductance L and resistance R. For the magnetic coil, the small induced potential will be neglected.

We wish to determine the dependence of the height $z(t)$ of the ball on the potential $u(t)$ at the magnet. For this purpose we are developing a block diagram with input signal $u(t)$ and output signal $z(t)$.

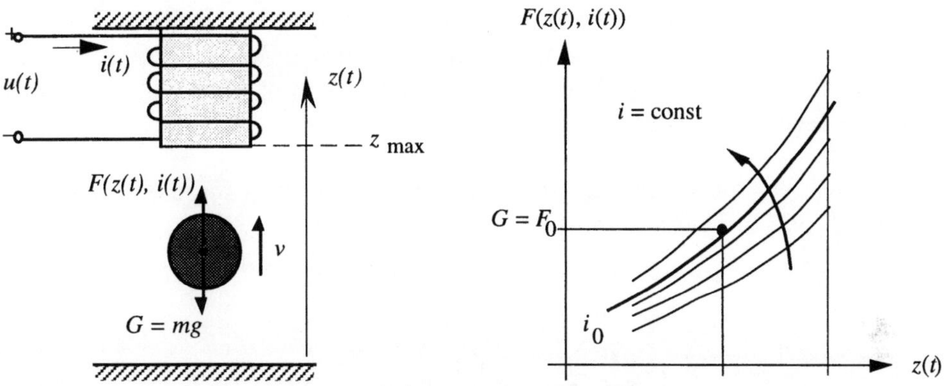

Figure 1. A ball suspended beneath a magnet with force operating characteristics

$u(t)$	=	magnet potential (input signal)
$i(t)$	=	magnet current
$z(t)$	=	height of sphere
$F(t)$	=	$F(z(t), i(t))$ = force operating characteristics
i_0, z_0	=	operating point values (steady state values)
z_{max}	=	maximum height of ball
$v(t)$	=	velocity of ball
m	=	mass of ball
G	=	weight of ball (gravity force)
g	=	acceleration due to gravity (9.81 m/s ^{2})

We obtain the following relationships for the system.

$$F(t) = F(z(t), i(t))$$

$$= F_0 + \frac{\partial F(z_0, i_0)}{\partial z}(z(t) - z_0) + \frac{\partial F(z_0, i_0)}{\partial i}(i(t) - i_0) \quad (1)$$

$$m\frac{d^2 z}{dt^2} = F(t) - G \quad (2)$$

$$u(t) = L\frac{di}{dt} + Ri(t) \quad (3)$$

Relationship (1) follows directly from the force operating characteristics, when a linearisation is carried out using the beginning of the Taylor series expansion of the function $F(z,i)$ at the point (z_0, i_0). Here the partial derivatives $\partial F/\partial z$ and $\partial F/\partial i$ are the slopes of the operating lines at the operating point (z_0, i_0). Relationship (2) represents Newton's law in the z-direction. Relationship (3) is the network equation for the magnetic coil with inductance L and ohmic resistance R. As was already mentioned, the induced potential (also known as armature feedback) has been neglected here.

Figure 2 shows the block diagram, one with the nonlinear force operating field $F = F(z,i)$ and one with the linearised operating field.

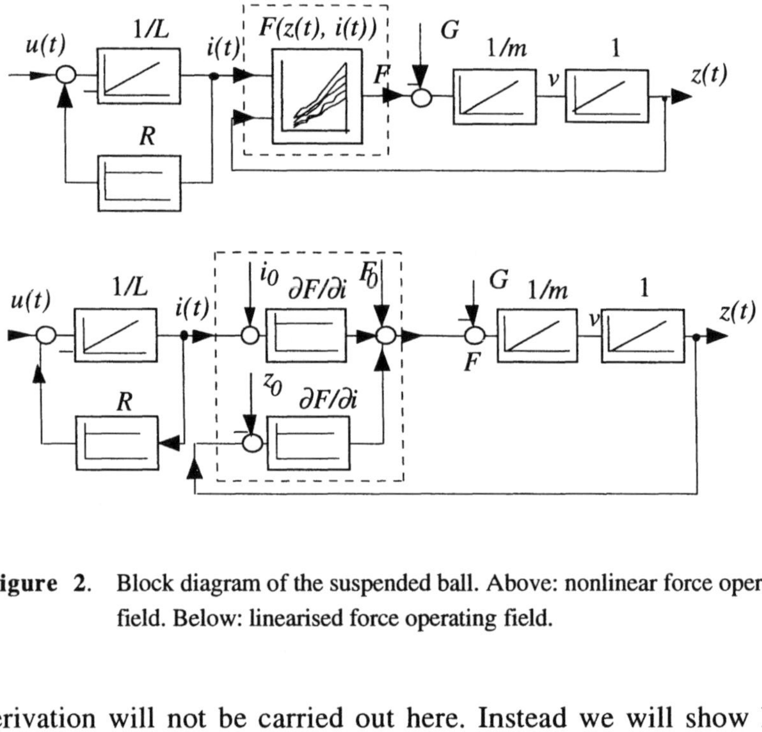

Figure 2. Block diagram of the suspended ball. Above: nonlinear force operating field. Below: linearised force operating field.

The derivation will not be carried out here. Instead we will show how the normalised block diagram can be derived.

Since we are only interested in the dynamics about the operating point (i_0, z_0), a normalisation will be carried out for this point. The following normalised deviations from the operating point are introduced:

height : $\Delta z(t) = \dfrac{z(t) - z_0}{z_0}$ current : $\Delta i(t) = \dfrac{i(t) - i_0}{i_0}$

force : $\Delta F(t) = \dfrac{F(t) - F_0}{F_0}$ voltage : $\Delta u(t) = \dfrac{u(t) - u_0}{u_0}$

In addition, we can define for convenience

magnet time constant $T_m = \dfrac{L}{R}$

normalised integration time constant $T_i^2 = \dfrac{z_0}{g}$

normalised operating line slopes $\dfrac{z_0}{G} \dfrac{\partial F(z_0, i_0)}{\partial z} = K_z$

$$\frac{i_0}{G} \frac{\partial F(z_0, i_0)}{\partial i} = K_i$$

At the operating point (i_0, z_0), the steady state point, it also holds that

$F_0 = G = mg$ (magnetic force = weight)

$u_0 = Ri_0$ (from (3) for $\dfrac{di}{dt} = 0$)

With these conventions a short intermediate calculation from equations (1) through (3) yields the normalised relationships:

$$\Delta F(t) = K_z \Delta z(t) + K_i \Delta i(t) \tag{4}$$

$$T_i^2 \Delta \dot{v}(t) = \Delta F(t) \tag{5}$$

$$\Delta u(t) = \Delta i(t) + T_m \frac{d\Delta i(t)}{dt} \tag{6}$$

Two tasks now lie before us:

- to construct the block diagram for the suspended ball (Fig. 3 shows the result),
- to determine the parameters for this diagram, when the device is specified.

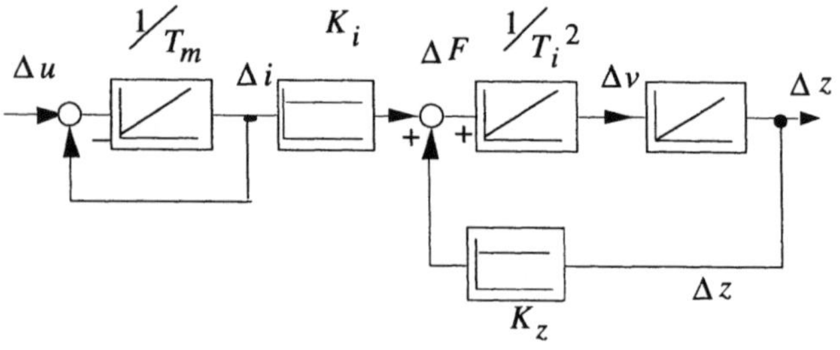

Figure 3. Normalised, linearised block diagram for the suspended ball

We first determine the model parameters (coefficients):

Ball mass: $m = 12.7g \rightarrow G = m \cdot g = 125\text{mN}$

Chosen operating point: $z_0 = 6$ mm, $i_0 = 280$ mA $\rightarrow F_0 = G = 125$ mN

Figure 4 shows the force operating line field for an actual device in the form of two families of curves. By drawing the tangents to these curves at the chosen operating point (z_0 , i_0), the two slopes at the operating point are determined.

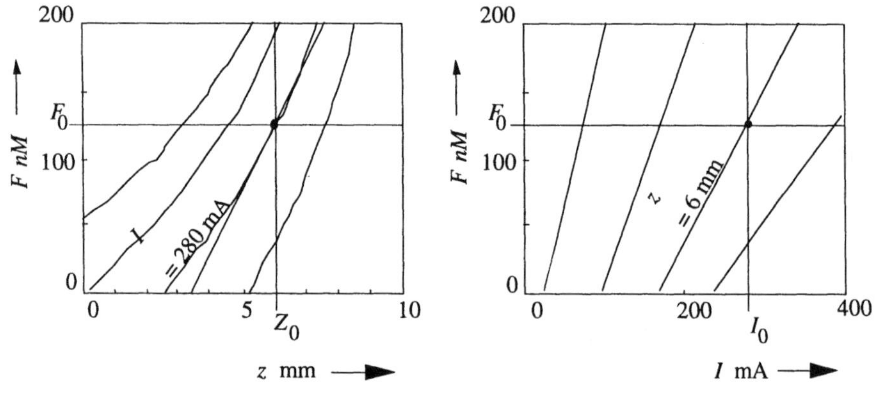

Figure 4. Measured force operating line field for a suspended ball

We can read off

$$\frac{\partial F(z_0, i_0)}{\partial X} = 47.0 \frac{\text{mN}}{\text{mm}} \quad \text{or} \quad \frac{\partial F(z_0, i_0)}{\partial i} = 1.15 \frac{\text{mN}}{\text{mA}}$$

When normalised this becomes

$$K_z = \frac{6}{125} \cdot 47.0 = 2.26$$

$$K_i = \frac{280}{125} \cdot 1.15 = 2.58$$

The normalised integration constant is then

$$T_i = \sqrt{\frac{z_0}{g}} = 24.7 \text{ ms}$$

$$L = 140 \text{ mH}, \ R = 10\ \Omega \ \rightarrow \ T_m = \frac{L}{R} = 14 \text{ ms}$$

Now we have determined all the parameters necessary for the block diagram in Fig. 3, which totally describes the system.

REFERENCES

Abel, H *et al* (1991). *Guidelines for Authors.* Tübingen: Deutsches Institut für Fernstudien an der Universität Tübingen.

Dörrscheidt, F and Latzel, W (1989). *Grundlagen der Regelungstechnik.* Stuttgart: Teubner.

Föllinger, O (1990). *Regelungstechnik.* Heidelberg: Hüthig.

Kull, H (1992). *Modeling of Transfer Systems. Computer Applications in Mechanical and Manufacturing Engineering, Course I, Module 3.* Ostfildern: Technische Akademie Esslingen.

Looze, D P (1992). *State Variable Models for Linear Time-Invariant Systems. Computer Applications in Mechanical and Manufacturing Engineering, Course I, Module 9.* Ostfildern: Technische Akademie Esslingen.

Looze, D P (1992). Control Synthesis in State Space. *Computer Applications in Mechanical and Manufacturing Engineering, Course I, Module 10.* Ostfildern: Technische Akademie Esslingen.

Ogata, K (1987). *Discrete-Time Control Systems.* Englewood Cliffs: Prentice Hall.

Reuter, M (1988). *Regelungstechnik für Ingenieure.* Braunschweig: Vieweg.

Unbehauen, H (1989). *Regelungstechnik 1, 2, 3.* Braunschweig: Vieweg.

Section 3

Looking ahead – research on mathematical modelling

18

Modelling intentions realised: assessing the full range of developed skills

Christopher Haines
City University, London, UK
John Izard
Australian Council for Educational Research, Hawthorn, Australia
David Le Masurier
University of Brighton, UK

SUMMARY

The place of assessment is considered in relation to the whole range of skills taught or acquired in a framework of mathematical modelling, and argues for the current view that this should be achieved by direct assessment rather than by traditional examinations. A research strategy is described and tested, which shows how the assessment was made more comprehensive in the tertiary sector in the UK. Data from the development and testing of both written and oral descriptors for mathematical modelling projects are used to highlight questions of comparability and accountability between examiners, projects and students.

1 THE ROLE OF ASSESSMENT

Where modelling is used, either in its own right or as a vehicle for the teaching of mathematics, a great many skills enter the learning environment of the student. Some of the skills acquired by the student will be additional to those which were included in the original aims and objectives of the particular course. There are compelling reasons that the full range of developed skills in the student, taught by modelling, is assessed. Before considering ways in which the assessment can be made comprehensive (in section 2), and the specific example of oral communication skills associated with mathematical modelling (in section 3), the underlying reasons that this should be done by *direct* means rather than using traditional methods are discussed.

Advocates of mathematics teaching by modelling activities, consider that these activities are more authentic, and address situations where there are many solutions, where there are many problem-solving approaches, or where the skills cannot be demonstrated easily in traditional examinations. Assessment can provide valid evidence of learning achievement which informs students, facilitates provision of further learning, or certifies that a required level has been reached. The connection between these functions is clear, and such information is of direct relevance to each student and every teacher (Clarke, 1989).

Examinations exert great pressure on the subjects taught, and on the teaching strategies used. The introduction of modelling into the curriculum, without changes in the method of examination is doomed to failure. If, however, changes are made in the type of examination too, then there will be consequential changes in content and method of education to improve the chances of success (Gasking, 1948). The effects of direct and indirect measurement of skills and knowledge have also been compared. Teachers will concentrate on those capacities measured directly in the examination, which may, in fact, be *indirect* measures of appropriate skills. This will divert education away from its stated objectives onto the measures used by the examiners. Traditional examinations avoid items which are difficult to measure, such as those which test understanding or use new applications. Good assessments show a high correlation between the intended skills and those which are actually measured, but this correlation is significantly reduced by coaching in the indirect criterion, so that this measure is then *not* a valid indicator of the true objectives. The situation is distorted even more when examiners see that students without the intended skills are passing and raise the standard set for the indirect criterion, making the examination an even-less valid measure of those skills (Gasking, 1948).

The only solution is to devise direct measures which are appropriate for the teaching of mathematics through modelling activities, so that the assessment strategy matches the curriculum intention. A comprehensive picture of students' learning achievements is provided by assessment strategies which cover knowledge, understanding, skills and personal qualities not assessed by traditional tests, such as the development of initiative, taking responsibility for learning, and applying problem-solving strategies. Use of mathematical projects as assessment tasks could ensure that the tasks mirror the desired skills, thus contributing to valid assessment. However the wide range of possible responses makes the use of such assessment strategies problematic, unless better procedures are devised, not least to comply with current academic audits (Haines, 1992).

The procedures need to deal with complexity, encourage better communication between examinee and examiner, and lead to more consistent and relevant assessment. The procedures should be feasible in current practice, and should indicate how confident the examiner is about the precision of the assessments. The assessment strategies adopted must meet accountability requirements – fairness to each student in respect of their genuine knowledge and skill, giving students the means to support their own learning, and providing certification with integrity.

To obtain assessments with these properties, common or agreed descriptors must be devised appropriate to the learning context. The descriptors should interpret the detailed aims and objectives of the exercise. Those who judge the projects should also share the same meanings for most of the descriptors. As projects may differ in the demands they make of candidates, choosing a project which makes less demand may give a student an advantage by allowing mastery of a topic to be demonstrated more easily. If each person does a different project, and no person does more than one project, then the influence of the project is inseparable from the achievement of the candidate. The assessment process adds its own problems. Examiners look for different features, emphasising them differently, have different views of complexity, and so assign differing values to the work and vary in their assigned marks. Examiners commonly partition the marks, awarding marks to sections of the work or categories of tasks (Haines, 1991 and 1992), but the nominal weightings given to the candidates may not be reflected in the marks actually awarded, distorting the intended balance of the assessment (Izard, 1991).

This paper describes research in which comprehensive descriptors of oral and written achievement were developed to assess complex behaviour (as in mathematical-modelling projects), to recognise high achievement, and to reward curriculum-intended achievement. Analyses of non-dichotomous data from trials of the scheme are presented, and interpreted in terms of the consistency and precision of assessing candidate achievement, the stringency of examiners, and the differing demands of course requirements.

2 DEVELOPING A COMPREHENSIVE ASSESSMENT STRATEGY

A research-assessment group of practitioners in the development of project schemes met in Exeter in 1991 to

- review current project schemes, and their assessment of project work, across a wide range of institutions and levels,
- begin to develop criteria-based assessment procedures for use on a wide range of topics within mathematics.

Schools, colleges, and universities were represented, ensuring that the projects which were available were completed within course frameworks which dealt with diverse aims and objectives. The projects brought to the workshop dealt with different topics, were completed at different ages, and demonstrated a variety of achievement levels. All this ensured that any assessment was not too narrowly focussed on content and year level, and that any process which rated projects could be applied to real differences in achievement (Berry and Haines, 1991).

Initial consideration was given to the descriptors (or indicators) which might be appropriate for objective assessments of projects. Two working groups began the task by reading the nine selected projects, to form an initial impression and to rank all nine pieces of work (their rankings are shown in Table 1). A second working session identified three main areas or themes.

- Activity of the investigation: processes in the constituent information structure.
- Integration: ability to integrate knowledge and skills to tackle a particular problem.
- Communication: delivery encompassing the written report, group work and oral presentation.

Descriptors of each of these themes were then developed, and rating sheets produced. The descriptors were phrases like 'fluent and effective use of language', and those using the rating scale were asked to assign a rating of 2, 1, or 0 to that descriptor according to whether the project being assessed showed comprehensive evidence, some evidence or no evidence, of this behaviour. An inappropriate category, rated *X*, was included for cases where the marker felt the descriptor to be inapplicable to that project. Six of the nine projects were remarked using the rating sheets, and a detailed item-response model analysis of this activity was carried out by Izard (Berry and Haines, 1991) using the FACETS routine described by Linacre (1990).

	Level	Group 1 rank	Group 2 rank	FACET rank
1 History	UG	6	5	2
2 Modelling 1	UG	3=	?	5
3 Modelling 2	UG	1	2	3=
4 Modelling 3	Dip	8	7	
5 Complex Analysis	UG	?	?	
6 Investigation 1	GCSE	3=	3	1
7 Investigation 2	GCSE	3=	1	3=
8 Investigation 3	UG	7	4	
9 Investigation 4	UG	2	6	6

Table 1. Initial ranking of specimen projects and ranking following FACET analysis. UG: Undergraduate, Dip: Diploma, GCSE: General Certificate of Secondary Education. (Berry and Haines, 1991).

FACETS is a computer package which uses multiple regression in its analysis, to provide a measure of student, judge and item (i.e project) performance in the assessment. In this research, the projects themselves were ranked on an achievement scale according to the ratings of the judges (see Fig. 1.) The term *judge* is used, to indicate a person who is faced with the task of assessing students work within the investigational framework of the research. The achievement scale is a measure in terms of standard deviations from the mean, either below or above. Figure 1 shows that projects 3, 7, 1 and 6 were above the mean, and within one standard deviation of the mean. The judge's performance is also interpreted relative to this scale (Fig. 2). Item-response theory is used to analyse the responses of judges to descriptors of judge and student behaviour.

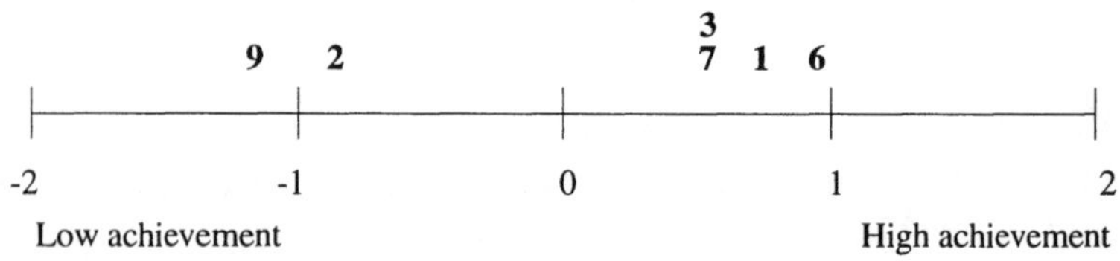

Figure 1. Projects ranked on an achievement scale.

There were three facets in the design of the analyses: Judges, Students/Projects, and Items (descriptors). The analyses used a Rating Scale Model – with allowed categories 0, 1, 2 – for all judges, for all persons/projects, and for all items with a weight of 1. That is, each judge was regarded as of equal importance, each student/project was regarded equally, and each of the 29 descriptors was regarded as of equal importance. The 'inappropriate' category, shown as *X* on the sheets, was excluded from the analyses. There were 676 usable ratings, defined as a 0, 1, or 2 rating on an item, by a judge, for a student/project.

There were significant differences between the achievement levels of students, as inferred from the evidence of the projects. Figure 1 represents the projects, on a linear scale, as a measure of achievement. The numerals above the axis are codes to identify the projects.

The rankings for the projects are given in Table 1, which shows that changing the assessment process through discussion, and concentrating on descriptors, changed the overall rating. Projects 9 and 2 were ranked some distance below the other projects on the achievement scale (Fig. 1), although it was not possible to determine a consensus position for a 'pass' mark since the 'pass' mark for each institution was not known.

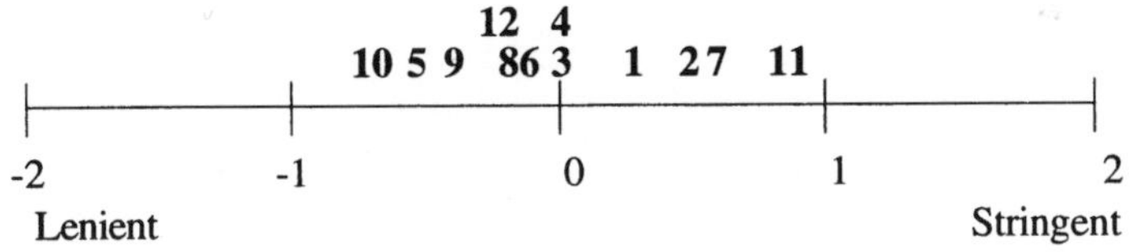

Figure 2. The judges, shown on leniency-stringency scale.

Figure 2 shows the judges, on a linear scale which could be described as a leniency-stringency dimension (in the same metric as for Fig. 1). The numerals immediately above the axis are codes to identify the judges. Note that all judges except judges 7 and 4 have similar patterns, but vary in their stringency. Judge 4 differs from the others by making less extreme distinctions between the projects; judge 7 differs by making more extreme distinctions between the projects. No inference should be made that one type of pattern is better than another. The remarkable consistency between judges (reliability 0.67), even though some are more lenient than others, is cause for celebration but not complacency. Developing strategies which improve such assessment of complex tasks remains an objective.

At first sight Table 1 highlights disagreement between the judges on the project assessment prior to the descriptors being developed, but Fig. 3 shows a reasonable correlation, +0.57, between the two groups of judges on the seven ranked cases. The individual behaviour, and the group behaviour of the judges, gives some ground for confidence in the quality of their assessment.

The 29 descriptors were analysed for consistency, and for the demands which they placed on the students. (The descriptors were placed on a continuous demand scale similar to Fig. 2). Of the three themes, the descriptors under 'Activity of the Investigation' included the extremes of demand.

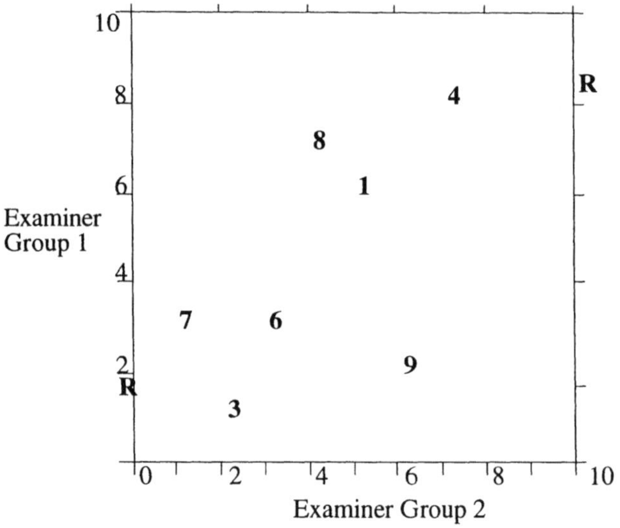

Figure 3. Assessment of written reports by two groups of examiners; project number plotted. The least-squares regression line joins the points marked R on the axes.

The descriptors

- identified the main features of the task
- relationships between variables established

were the easiest to satisfy, whilst

- hypothesis stated and established
- good use of human resources made

were the indicators on which it was most difficult to achieve a good score. In addition, redundancy was detected, under the same theme, between

- results interpreted and validated
- conjectures analysed

whilst the model did not fit

- generalisations made and proofs attempted

which suggests that this descriptor should have been separated into two distinct items.

The workshop explored the central issues behind marking projects, and suggested categories and indicators, with associated rating sheets, which might be used for this purpose.

Further data were collected following the Exeter workshop (Izard and Haines, 1991) in order to develop and test these strategies for producing effective rating scales in the tertiary sector. As part of this exercise, the projects listed in Table 1 were assessed by some of the same judges, using a larger collection of descriptors, which were not clustered into categories. The assessment process using two different sets of descriptors gave almost identical results, with a correlation of 0.92 (Fig. 4). We consider that this reflects the importance of the judges sharing the same meanings, rather than attempting to find a unique set of descriptors.

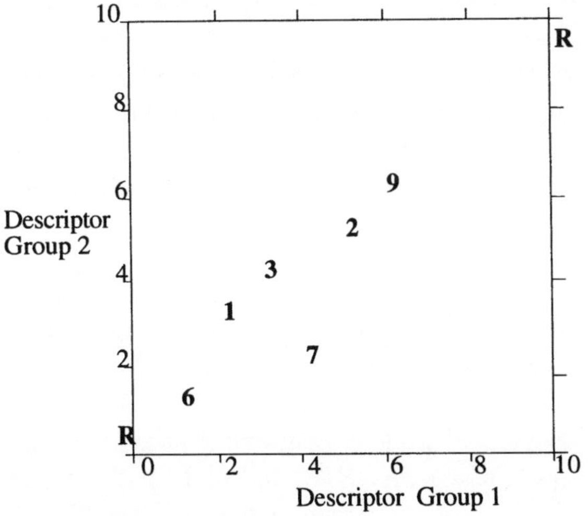

Figure 4. Rank assessment of written reports from two sets of descriptors; project number plotted (so that project 7 was ranked 4 using the descriptor group 1 and ranked 2 using descriptor group 2). The least-squares regression line joins the points marked R on the axes.

3 DEVELOPING THE ORAL DESCRIPTORS

Oral communication descriptors developed at Exeter (Berry and Haines, 1991) were tested in a pilot investigation alongside the official assessment procedure at Brighton University.

In the first stage, two judges took the thirteen descriptors to the group presentations of a modelling project on insurance. Difficulties were found with the assigning of individual and group measures of performance – for instance it was not clear that, referring to visual aids, the term 'well explained' was or was not an individual attribute. The data obtained from this exercise were too sparse and inconsistent to give any reliable indication of performance.

In the second stage, the same descriptors were used in group presentations given, by the same students, on a different modelling project. Of the four judges, two were internal examiners. The topic and the assessment was the same for all groups, but there was no requirement that the presentation of the report should be handled in the same way. A video of the presentations was made to allow further follow-up analysis.

At group level each judge ranked the 10 groups by reference to the communication skills guidelines, but not using a quantitative measure. A FACET analysis was carried out on the data provided by three of the four judges, using the descriptors. The data provided by the fourth judge were very sparse, and so were omitted as atypical. Although the global assessment marks correlated with scores obtained using the FACET analysis, there were substantial discrepancies at the extremes, giving rise to the inference that the judges did not have shared meanings for the descriptors – they did not describe those qualities which the internal judges held to be important, at least where work of the highest and lowest standards were concerned.

Turning now to individual student assessments, the responses from three of the judges were analysed using FACETS, and their performance compared. It can be seen that there was some agreement between judge 1 and 2 (correlation 0.46, Fig. 5), but a comparison between judge 1 and judge 3 showed a lower correlation (0.11) as did that for judge 2 and judge 3 (0.16).

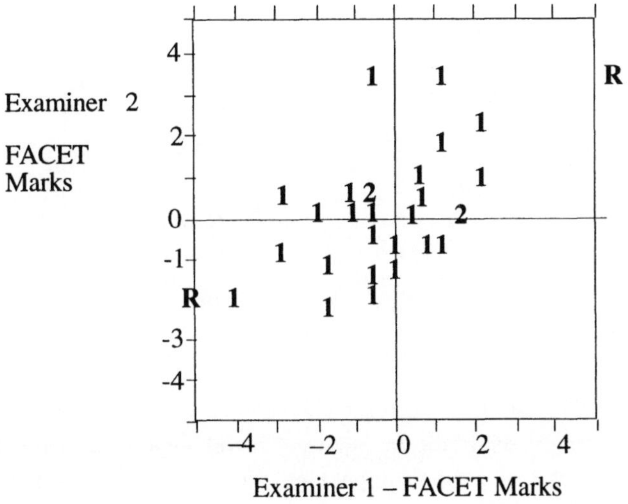

Figure 5. Assessment of individual oral reports; judge 2 compared with judge 1, correlation 0.46. 28 students are plotted. Examiner 1 scored the worst student -4.2 on an assessment scale whilst the same student was scored -2.5 by examiner 2.

The internal-examination scheme required that each student be graded fail, pass, merit or distinction; these grades were coded 0, 1, 2 and 3 respectively. The grades awarded by the internal examiners are compared with those which would have been awarded had the FACET analysis been used on descriptors of performance for each individual student (Fig. 6). The vertical scale in Fig. 6 shows the students clustered at the four levels according to the overall assessments of the internal examiners, and the horizontal scale shows each student rated on an ability scale using the FACET analysis. Whilst there is a positive correlation (0.71) between the two measures, there are substantial discrepancies which give cause for concern. At the right hand side of Fig. 6, there are three students who appear to be candidates for a distinction, one of whom achieved a bare pass. There is evidence of a considerable overlap in the merit and pass categories, which suggests that the internal criteria were not well-defined and/or that the descriptors either do not measure student achievement as intended or that the judges do not share the same meanings for the descriptors.

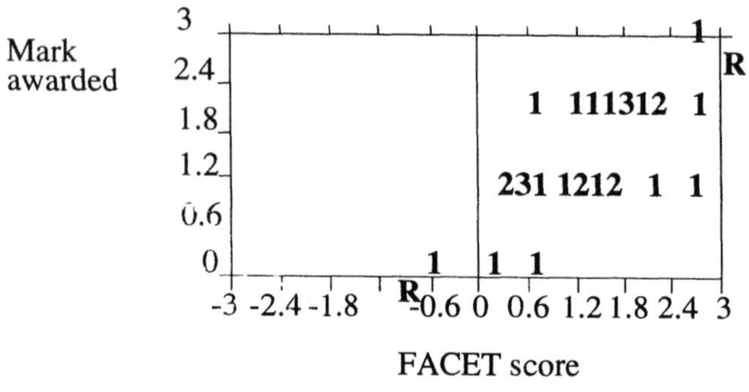

Figure 6. Individual student performance in oral presentations, 28 students are plotted. Comparison of the grades awarded with the FACET score. (The 3 students on the right of the figure have the same score 2.4, but the internal examiners awarded marks 1 (pass), 2 (merit) and 3 (distinction)).

4 POSTSCRIPT

Item-response theory is a powerful tool in the quest for credible assessment schemes for mathematical modelling and other projects. The foregoing analysis shows that, in developing such schemes, thorough testing and research is necessary. The oral descriptors tested above were shown to have inherent difficulties in reflecting the perceived curriculum aims, and in the judges achieving shared meanings prior to use. A subsequent workshop early in 1992 made further progress in developing a robust set of oral descriptors – see Appendix – which is now undergoing further testing in the UK.

A five-point scale from high achievement (4) to low achievement (1), the fifth point registering 'not shown', was used rather than a three- or four-point scale, since it leads to higher consistency between results and consequent increased reliability. The accompanying commentary (see Appendix) indicates the basis on which the judgement is to be made. A given institution or course could construct their own assessment form, with the commentary being used as an aide memoir. The map from the rating scales to grades or marks can be developed, to suit the institution, by group discussion of typical exemplars.

At the same time, descriptors which assist in the measurement of skills associated with content and written communication are being developed in a form which allows wide applicability. The central questions relating to learning mathematics through project work remain. It is important that the intended skills are identified in the descriptors, and that the assessment process does not distort the curriculum intentions.

REFERENCES

Berry, J S and Haines, C R (1991). *Criteria and Assessment Procedures for Projects in Mathematics.* Workshop report, Exeter, 15-17 April 1991. Plymouth: PolySouthWest.

Clarke, D J (1989). *Mathematics Curriculum and Teaching Program (MCTP): Alternatives in Mathematics.* Canberra: Curriculum Development Centre.

Gasking, D A T (1948). *Examinations and the Aims of Education.* Carlton: Melbourne University Press.

Haines, C R (1991). Project assessment for mathematicians. In Niss, M, Blum, W and Huntley, I (ed), *Teaching of Mathematical Modelling and Applications.* Chichester: Ellis Horwood, 209-305.

Haines, C R (1992). Developing assessment strategies for mathematics projects. In Stephens, M and Izard, J (ed), *Reshaping Assessment Practices: Assessment in the Mathematical Sciences under Challenge.* Hawthorn: Australian Council for Educational Research, 127-141.

Izard, J (1991. Issues in the assessment of non-objective and objective examination tasks. In Luitjen, A J M (ed), *Issues in Public Examinations.* Proceedings of the Sixteenth IAEA Conference, Maastricht, 18-22 June 1990. Utrecht: Lemma, BV, 73-83.

Izard, J and Haines, C R (1991). Marking Schemes for Projects in Mathematics and its Applications: Interpreting the Results. Paper presented at ICTMA-5, Utrecht, 9-13 September 1991.

Linacre, J M (1990). Modelling Rating Scales. Paper presented at the Annual Meeting of the American Educational Research Association, Boston, 16-20 April 1990. [ED 318 803]

APPENDIX

Communication skills (oral)

	high			low	not shown
O1 Rapport with the audience	☐	☐	☐	☐	☐
O2 Effective delivery	☐	☐	☐	☐	☐
O3 Command of spoken English	☐	☐	☐	☐	☐
O4 The structure gets over their main points	☐	☐	☐	☐	☐
O5 Clear explanation of problem and outcome	☐	☐	☐	☐	☐
O6 Overall planning and organisation	☐	☐	☐	☐	☐
O7 Appropriate use of visual and other aids	☐	☐	☐	☐	☐
O8 Technical quality of visual and other aids	☐	☐	☐	☐	☐

Commentary

Descriptors O1 - O4 are for individuals in a group *or* for individuals making an individual report.
Descriptors O5 - O8 are group descriptors for the whole group, *or* additional descriptors for individuals making an individual report.

O1 Engaging and keeping the interest of the audience; eye contact; enthusiasm; personal qualities etc.

O2 Speed, pace, tone, pauses, intonation, well rehearsed, spoken not read, rhythmn.

O3 Sentence structure, vocabulary, grammar, appropriate use of appropriate language.

O4 Explained well, evidence of structure, appropriate reference to mathematics, understanding.

O5 Clear explanation of the problem, effective presentation for audience.

O6 Evidence of a plan, organisation (at group level), work by a team, coherence.

O7 Choice of the right medium for communication and good use of it. Aids present when necessary. Explanation.

O8 Aids produced with clarity, well set out and labelled.

19

Reflections on future developments in the light of empirical research

Gabriele Kaiser-Messmer
University of Kassel, Germany

SUMMARY

On the basis of the results of comparative empirical research between English and German schools proposals are developed in order to improve the unsatisfactory situation concerning the inclusion of real-world and modelling examples in everyday school practice. The following issues are tackled: the participation of the students in the modelling process; the balance between informal, context-centred approaches and more formal, mathematically-oriented approaches within the concept introduction; the authenticity of the real-world examples included in the mathematics lessons; the modes of assessment; the influence of technology on concept comprehension.

1 INTRODUCTION

Many reports, from different countries around the world, state that real-world and modelling examples do not have the importance in the classroom given them in the didactical discussion. In many countries they still do not play an important, indispensable part in mathematics teaching. Furthermore, even in countries in which real-world and modelling examples are included in the curricula of many schools, the examples are very often restricted to small, artificial problems allowing only a limited range of activities for students. In the following, I will develop a few proposals for educational measures in order to improve this situation, which are based on comparative empirical casestudies that have recently been carried out in English and German schools, in the context of a joint collaborative project between Exeter University and Kassel University. For a better understanding of the proposals developed, the teaching approaches and the materials on which the casestudies were based will be described.

So far in this research, several casestudies in English and German school classes (students aged 14-16) have been carried out based on teaching materials which are characteristic either for the English or the German application-oriented mathematics education. As a research method, the techniques of classroom observations, attitude tests, and achievement tests have been used. The casestudies in the English schools were based on an experimental course for the teaching of mathematics in context, the Enterprising Mathematics Course (Hobbs, 1992, 1993). This course is typical of recent English experimental approaches to the teaching of mathematics in context, because of its structure around real-world problems, its strong emphasis on the development of mathematical concepts out of real-world contexts, and concept introductions restricted to contextual reflections. Students' activities were usually centred around coursework, which is a special kind of continuous-assessment mode common, but not compulsory, in the last two years of the lower-secondary level in English schools. Here the students worked independently over a period, at home and in school, in order to solve a mathematical or real-world problem, which was usually selected by the students themselves. The casestudies in the German schools were either based on an application-oriented textbook (Mathematik heute) or materials of the Mathematik-Unterrichts-Einheiten-Datei. These materials were typical of the German approach which means, among other things, the sequencing of the course took place along an explicit mathematical structure, the mathematical concepts were developed within real-world contexts with an emphasis on different basic mathematical ideas, and the concept introductions were more formal and precise compared with those in the English course. In German schools there is no equivalent to coursework; instead assessment was done by time-limited teacher-developed written tests (for a detailed description of the casestudies see Blum *et al*, 1992).

On the basis of these empirical studies, I feel that the following aspects should be considered in order to improve the teaching and learning of real-world and modelling examples.

2 ACTIVE ENGAGEMENT IN MODELLING WORK

It is necessary to have extensive periods in which the students are actively engaged in modelling work, and are able to work through all phases of the modelling process for themselves.

The casestudies show that the inclusion of real-world problems into mathematics teaching without the active participation of the students does not promote the development of the abilities to apply mathematics in unknown real-world examples or to carry out the modelling process. The English, as well as the German, students were only able to solve real-world examples similar to the ones they had dealt with in previous lessons. With unknown real-world situations in which they were only asked for mathematisations, they had severe difficulties. This observation is supported by the results of earlier research related to modelling courses (such as by Treilibs, 1979), which emphasise the necessity of students' own activities. The underlying assumption of this plea for active modelling work of the students may be described as a transfer hypothesis – it is expected that students who have actively been involved in modelling processes have learned to use mathematics in unknown situations. Further empirical studies are necessary to substantiate this hypothesis.

3 CONCEPT INTRODUCTION

The concept introduction has to find a balance between informal, context-centred approaches and more formal, mathematically-oriented approaches.

The English experimental course observed mainly utilised a context-centred concept introduction, without further generalisations, which makes use of these concepts in other contexts more difficult. Without further assistance the students had severe difficulties in transferring example-centred definitions to a generalised comprehension. Furthermore, many students did not gain insight into theoretical relationships. On the other hand, the German method of concept introduction, which was very often more formal and abstract mathematically, caused severe difficulties with the retention of these concepts, especially for weaker students who sometimes did not even remember the basic ideas.

4 AUTHENTICITY OF EXTRA-MATHEMATICAL PROBLEMS

A significant number of extra-mathematical problems included in the mathematics lessons have to be authentic, in order to convince the students that mathematics is really used in their lives.

Referring to a description given by Niss (1992), I define an authentic extra-mathematical situation as one which is embedded in an existing practice or subject area outside mathematics, and deals with objects, phenomena and issues that are genuine to that area and are recognised as such by people in it. This definition does not necessarily require that the extra-mathematical situation has to deal with everyday practical matters from people's lives. I think, however, it is necessary that a few of the extra-mathematical problems treated in the mathematics lessons are relevant to the actual lives of the students, and are seen in this way by the students.

The casestudies we have carried out show that, even after a diet of application-oriented teaching, it is not self-evident to the students that mathematics is connected to their everyday life. As illustration I give below a few typical answers, of both English and German students, from the casestudies. The students were interviewed after several months of application-oriented mathematics teaching, and asked whether they had ever used the mathematical methods learned in school in everyday life or other school subjects:

"Yes. I have learnt to check my change" or "I counted the amount of calories in my daily intake" or "Yes, when money is involved and also helping my brother do homework".

Generally the descriptions of the situations where they had used mathematics, given by the English and the German students, were rather poor, and showed that they either did not use the mathematics learned in school or were not aware of it. The kind of research described in this paper, which is based on written interviews with the students, gives no evidence whether the students have gained a better understanding of real-world phenomena without being conscious of it. Results from ethnographical studies point out that, especially in non-western countries, a lot of mathematics is involved in the everyday life of many people without being recognised as mathematics (see, amongst others, Rogoff and Lave, 1984). So far, no attempts have been made to transfer these results into the context of western school systems.

5 NEW MODES OF ASSESSMENT

There is a consensus that new modes of assessing modelling examples are necessary. The instrument of coursework seems to be one of several adequate instruments to assess modelling.

Coursework allows the students to carry out the entire modelling process, assisted by the teacher when necessary. The students are encouraged to find their own problems, and to see mathematical questions in ordinary real-world problems. I am, of course, aware of the many difficulties which arise with coursework – the difficulties in finding appropriate problems and themes, which can be tackled in an appropriate amount of time, which are appealing to the students, can be assessed in a less subjective way, and so on (see for example Jennings, 1992). On the basis of the examples of coursework in our casestudies, I claim that good organised coursework combines the assessment task with the possibility of the promotion of independent, active work by the students.

6 BARRIERS BUILT UP BY TECHNOLOGY

The use of technology, which will significantly change the mathematics curriculum in the future, has advantages, but can sometimes build up barriers for an adequate concept comprehension as well.

In the casestudies, many English students had severe difficulties with rearranging formulae – which was caused by experience of notations, contradictory to mathematical conventions, used in pocket calculators. Several students have been accustomed to notation styles such as the following.

$$r = \sqrt{V \div h \div \pi} \qquad r = V \div h \div \pi\sqrt{} \qquad \frac{V}{h} \div 3.14\sqrt{} = r$$

I am convinced that, despite this rather negative description of the topical situation, it will improve in the near future – there are many teachers and educators all over the world who are aware of these problem areas, and have started various educational initiatives in order to change this situation.

REFERENCES

Blum, W *et al* (1992). Teaching and learning of mathematics and its applications: first results from a comparative empirical study in England and Germany. In *Teaching Mathematics and its Applications*, **11**, 112-123.

Hobbs, D (ed) (1992, 1993). *Enterprising Mathematics Course*. Oxford: Heinemann.

Jennings, S (1992). A Review of Coursework in Mathematics at GCSE. In *Teaching Mathematics and its Applications*, **11**, 64-70.

Niss, M (1992). *Applications and Modelling in School Mathematics - Directions for Future Development.* Roskilde: IMFUFA Roskilde Universitetscenter.

Rogoff, B and Lave, J (ed) (1984). *Everyday Cognition: Its Development in Social Context.* Cambridge: Harvard University Press.

Treilibs, V (1979). *Formulation Processes in Mathematical Modelling*. Nottingham: Shell Centre for Mathematical Education.

20

Necessary research in mathematical modelling and applications

João Pedro da Ponte
University of Lisbon, Portugal

SUMMARY

Curriculum development in mathematical modelling and applications is an active area of work in mathematics education. However, it needs to be more closely connected to research. This paper suggests five different areas in which research may be important: epistemological, cultural and social issues; students' cognitive processes; curriculum approaches; didactical strategies and classroom processes; teacher development. Some examples of recent and ongoing studies are provided. The paper concludes by presenting a few remarks about the nature of the necessary research, and sketching some ways in which this effort can be undertaken.

1 INTRODUCTION: NECESSARY RESEARCH?

Just by itself, or in association with computers, mathematics plays an increasing rôle in modern society. It is used to convey ideas, to support controversial arguments, to control information flows, and to assist in problem solving. For such purposes it may be viewed as a descriptive language, a classificatory instrument, an encoding mechanism, or a decision-making tool. This implies that the ability to use mathematics in concrete situations, and to analyse critically how it has been used by others, become in themselves important educational objectives.

In most countries, applications of mathematics and modelling have a secondary, if not negligible, place in the school curriculum. To reverse this situation, it will be necessary – amongst other things – to undertake considerable work in terms of curriculum development and research. Curriculum development includes the production of well-articulated proposals, supporting materials, classroom activities, teaching modules, and integrated courses aimed at these educational objectives. Research needs to address all the problems that emerge in implementing these activities at the national, school, and classroom level, and to suggest strategies for overcoming the inevitable difficulties.

Curriculum development is already taking place for in several countries, and there is a considerable bibliography in this area (see Blum, 1994). However, there are some inherent weaknesses in this effort – a critical one concerns its insufficient research base. This paper indicates some of the areas in which research is most clearly needed. When appropriate, examples of recent or ongoing studies are provided.

2 GENERAL AND EPISTEMOLOGICAL ISSUES

One important area for research is the nature of mathematical models and the epistemological, cultural and social status. There are models with quite diverse characteristics, purposes and functions; however, there is a shortage of literature that adequately deals with this matter.

The significant work of Davis (1988), Skovsmose (1988), and Booss-Bavnbek (1991) point to some critical features of misusing mathematical models. Other kinds of analysis, emphasising historical and mathematical dimensions, are necessary to yield constructive orientations for curriculum decision-making and for classroom teaching.

Consider, for example, the common occurrence of dynamic systems in the educational literature on modelling. Indisputably, they have very interesting properties and find many applications, but that is not a reason to downplay the importance of other kinds of models. We need, therefore, a comprehensive taxonomy of models, taking into account their nature and relevance, enabling a better informed choice of curriculum options and of the corresponding implications.

3 STUDENTS' COGNITIVE PROCESSES

The cognitive processes of students, when they work in modelling activities and in applications of mathematics, constitute an important area of research which is already in progress.

Boero (1992), drawing on extended experience from a long-term development project, claims that *complexity* in applied mathematical problems concerns three distinct aspects,

- the number of choices and steps that the pupil must keep under control,
- the coordination of different concepts and procedures,
- the selection of pertinent aspects in a 'complex' reality.

A framework for the cognitive processes involved in model formulation was proposed by Lambert, Steward, Manklelow, and Robson (1989). It was set at three levels,

- beliefs, including other knowledge and fundamental assumptions,
- metacognition, that is monitoring, evaluation, planning and control,
- reasoning within the mathematical domain and reasoning within the problem domain.

However, further work is necessary to ascertain the relationships between all these elements.

Lesh (1982) and Lesh and Kaput (1988) also produced an interesting theorisation about such processes, based in what is called 'local conceptual models'. According to this, related questions are

- to identify the nature of the primitive conceptual models of the students for different mathematical ideas,
- to explain how the deficiencies of the diverse models are detected,
- to explain how the most successful models are gradually constructed,
- to describe the cognitive characteristics associated with the use of unstable conceptual models, and how their negative influences can be minimised.

This research concerns students' work on problems that are interesting and realistic, but mathematically quite simple. Will such observations have similar importance for other kinds of problems?

Reasoning in situations involving data gathering as part of the modelling process was investigated by Hancock and Kaput (1990). They contend that, for young students the iterative process of data modelling is essentially "a tangle of loops small and large", whereas for experienced data modellers it comes closer to "a grand loop from data design to data collection to data analysis and back to the beginning". This is also a line of work that deserves to be continued.

There are many other issues relative to cognitive processes that need to be considered.

- In what specific aspects are the processes used by the students in applied situations different from those that they use in current mathematical problems?

- What are their greatest difficulties in dealing with applied situations? In what way will their general conceptions and attitudes towards mathematics tend to interfere with the realisation of modelling activities?

- Can the students adopt the modelling cycle (Kerr and Maki, 1979; Niss, 1989) as a fundamental conceptual reference? In what parts of this cycle have they greatest difficulties? What heuristics relative to the global process, or relative to each phase may assist the students in solving applied mathematical problems?

4 CURRICULUM

We also need research on the curriculum itself. The introduction of applied-mathematical problems and modelling tasks presupposes a clear indication of objectives and strategies, and an articulation with the remaining content, objectives, and curriculum activities. As this is virtually a new strand to be introduced in the teaching-learning process, we need to study the related implementation problems carefully.

Modelling and applications may appear in the curriculum in several ways. They may constitute

- the fundamental subject, which gives the structure to the course (as suggested by Garfunkel, 1994),

- activities that run in parallel with a programme organised around established mathematical topics,
- mostly starting points for the introduction of new topics.

The curriculum may stress

- extended projects, that take weeks or months to complete,
- situations that may require one or two classes,
- more specific activities, several of which can be completed in a single class-period.

Furthermore the curriculum may emphasise the construction of models, or the use and exploration of models already constructed. It may, finally, consider situations taken from physics, biology, economy, or ones from modern life that are not clearly related to any specific subject matter area.

Van Streun (1990) compared three kinds of curriculum, all giving importance to applied problems. The first offered just mathematical and applied problems, the second gave implicit attention to heuristic methods and little formulation of mathematical concepts and techniques, and the third gave explicit attention to heuristic methods and more attention to mathematical concepts and procedures. He concluded that this last approach led to superior cognitive schema.

With a different focus, some important work in this area was also carried out by Kaiser-Messmer (1991), who indicated different orientations that may inform a curriculum

- stress in utilitarian and pragmatic activities,
- emphasis in scientific and humanistic educational goals,
- an integrated approach.

This kind of work must be continued, contrasting experiences realised in different countries, and discussing questions such as the following.

- What are the implication of each curriculum option? In which circumstances do they work satisfactorily?

- How to articulate the use of real-life situations as a starting point for concept learning (and learning of mathematics content in general), and as problem situations that must be studied in depth in themselves?

- What is the didactic rôle of the more ambitious projects and of the shorter activities? When must each one be used, and in what conditions?

- How do we establish different kinds of real-life situations, related to different underlying models, with different pedagogical rôles?

5 DIDACTICAL STRATEGIES AND CLASSROOM PROCESSES

We now have the question of didactical strategies at the classroom level. Broad curricular orientations still give room for very different possibilities – some very likely better than others – when one comes to the teacher-student interaction. Unfortunately, research conducted in real classroom settings tends to be very scarce – for an exception, see Abrantes (1991).

To understand the difficulties of the teachers, and to be able to give them adequate support, this is a most necessary area of work. Significant questions are, for example, the following.

- How to help the students to become more competent in solving applied mathematical problems? In what ways may the students be supported by the teacher? What pedagogical principles must be considered? What are adequate forms of intervention?

- What grouping strategies are more adequate for each kind of classroom activity – individual work, group work, mixed?

- What tasks must be done in the classroom, and what must be done outside the classroom – and how?

One further question, that inevitably arises in any curricular and didactical option that embodies a truly innovative approach, is how to carry out the assessment of the students (de Lange, 1987; Leal, 1992).

5 TEACHER DEVELOPMENT

Finally, it is necessary to undertake research about the teachers and the teachers' development processes. Teachers are not just obstacles to innovations. They have their own rationales, constraints, and resources. They come into teaching with distinct training, motivations, and competence. With respect to the relation of mathematics and reality we need to address questions such as the following.

- What difficulties have teachers in understanding the nature and rôle of applications and modelling?
- What difficulties do they have in terms of carrying out these activities in their educational practice?
- How do teachers tend to use different sorts of curriculum materials to support applications and modelling?
- What kinds of professional development programmes may be developed to help overcome their difficulties and insecurity?

Very often primary and secondary school teachers, and others as well, think that research has very little to do with them – it is something to be carried out by others, in higher education. That is a too restrictive view of research, that pays lip service to the needs of the teachers themselves. I would like to propose that there should be a close connection between curriculum development and research, or at least some kinds of research.

7 CONCLUSION

All these research areas are closely interrelated, interacting with each other. All of them need an urgent attack. We must have theoretical analyses – to clarify concepts and define fundamental questions, to help to organise conceptually this still messy domain. We also need empirical research – that contrasts the different proposals with the realities of practice, and helps to distinguish what works (at least in each particular case) from what looks interesting but does not work so well.

Research, here, is essentially the search for answers to well-posed questions by rigorous, systematic and verifiable processes. There are certain kinds of research that require a very heavy instrumental and conceptual machinery. Of course, some attention is needed not to vulgarise the term research, as is sometimes done. However, it should be recognised that there are simpler forms of research, which imply far fewer resources and which and are still suitable to address the problems at hand. At any rate, much is to be gained if we can connect more closely research with development, and school teachers with researchers and teacher educators, in the frame of well-established projects.

REFERENCES

Abrantes, P (1993). Project work in school mathematics. In de Lange, J *et al* (ed), *Innovation in Maths Education by Modelling and Applications.* Chichester: Ellis Horwood, 355-364.

Blum, W (1994). Mathematical modelling in mathematics education and instruction. In Breiteig, T, Huntley, I and Kaiser-Messmer, G (ed), *Teaching and Learning Mathematics in Context.* Chichester: Ellis Horwood, 1-12.

Boero, P (1992). The crucial role of semantic fields in the development of problem solving skills in the school environment. In Ponte, J P *et al* (ed), *Mathematical Problem Solving and New Information Technologies: Research in Contexts of Practice.* Berlin: Springer-Verlag, 77-91.

Booss-Bavnbek, B (1991). Against ill-founded, irresponsible modelling. In Niss, M, Blum, W and Huntley, I (ed), *Teaching of Mathematical Modelling and Applications.* Chichester: Ellis Horwood, 70-82.

Davis, P J (1988). Applied mathematics as social contract. In *Zentralblatt für Didaktik der Mathematik,* **20**, 10-15.

Garfunkel, S (1992). Future of mathematical modelling in the classroom. In Breiteig, T, Huntley, I and Kaiser-Messmer, G (ed), *Teaching and Learning Mathematics in Context.* Chichester: Ellis Horwood, 234-242.

Hancock, C and Kaput, J (1990). Computerised tools and the process of data modelling. In Booker, G, Cobb, P and Mendicuti, T (ed), *Proceedings of PME 14.* Vol 3. Mexico, 165-172.

Kaiser-Messmer, G (1991). Application-oriented mathematics teaching: a survey of the theoretical debate. In Niss, M, Blum, W and Huntley, I (ed), *Teaching of Mathematical Modelling and Applications.* Chichester: Ellis Horwood, 83-92.

Kerr, D R and Maki, D (1979). Mathematical models to provide applications in the classroom. In Sharron, S and Reys, R (ed), *Applications in School Mathematics.* Reston: NCTM, 1-7.

Lambert, P *et al* (1989). A cognitive psychology approach to model formulation in mathematical modelling. In Blum *et al* (ed), *Applications and Modelling in Learning and Teaching Mathematics.* Chichester: Ellis Horwood, 92-97.

de Lange, J (1987). *Mathematics, Insight and Meaning.* Utrecht: OW&OC.

Leal, L (1992). *Avaliação da Aprendizagem num Contexto de Inovação Curricular.* Master thesis, University of Lisbon.

Lesh, R (1982). Modelling students' modelling behaviour. *Proceedings of the Fourth Annual Meeting of the North-American Chapter of PME.* Athens, 181-188.

Lesh, R and Kaput, J (1988). Interpreting modelling as local conceptual development. In de Lange, J and Dooman, M (ed), *Senior Mathematics Education.* Utrecht: OW & OC, 60-68.

Niss, M (1989). Aims and scope of applications and modelling in mathematics curricula. In Blum, W *et al* (ed), *Applications and Modelling in Learning and Teaching Mathematics.* Chichester: Ellis Horwood, 22-31.

Skovsmose, O (1988). Mathematics as part of technology. In *Educational Studies in Mathematics*, **19,** 23-41.

van Streun, A (1990). The teaching experiment "heuristic mathematics education". In Booker, G, Cobb, P and Mendicuti, T (ed), *Proceedings of PME 14.* Vol 1. Mexico, 93-100.

21

Developing metacognition through peer and self assessment

Howard Tanner and Sonia Jones
Department of Education, University College of Swansea, UK

SUMMARY

This paper describes an action research project into the use and practical application of mathematics. It is claimed that, as part of learning mathematical modelling, students have to be 'socialised' into a wider view of mathematics. Assessment is an integral part of this process. Peer and self-assessment techniques provide for the explicit development of the metacognitive skills – planning, monitoring and evaluating – which are essential for modelling. Through reflecting on and assessing the work of others, students learn to regulate their own thinking. Whilst learning the skills of self-assessment students are learning how to model.

1 METACOGNITION AND MODELLING

There is increasing recognition of the importance of the rôle played by metacognition in mathematical problem solving and modelling (Dreyfus, 1990).

It has been argued that explanations of problem-solving diffculties based on purely cognitive factors are incomplete (Lester, 1987), and attempts have been made to integrate metacognition into theories of modelling and problem solving (Schoenfeld, 1987).

Metacognition involves awareness and control of one's own thinking (Brown, 1987). Distinct elements in metacognition have been identified, including knowledge of one's own thought processes, the control and application of that knowledge, and the learners' beliefs about the nature of mathematics and themselves as learners (Schoenfeld, 1985; Campione, 1987; Gray, 1991).

Research has shown that knowledge alone is not sufficient for successful modelling. The student must also choose to use that knowledge, and to monitor the progress being made (Silver, 1987; Alston and Maher, 1987; André, 1989). Whilst comparing expert and novice problem solvers, Schoenfeld (1987) found that the experts continually monitored their use of strategies and evaluated their progress. Novices, although they knew sufficient mathematics to solve the problem, never thought to use that knowledge. They failed to monitor and control their own thinking.

Gray (1991) identified three strands in metacognition which support the modelling process – planning, monitoring and evaluating. During modelling mathematicians alternate between these strands, continually suggesting ideas and strategies, evaluating and criticising them, and monitoring the progress made. That is, expert modellers are able to discuss and argue within themselves (Schoenfeld, 1987; Wheatley, 1991). Students need to develop such metacognitive skills of 'inner speech' in order to become effective modellers.

Hirabayashi and Shigematsu (1987) argued that students develop their concepts of metacognition by copying their teacher's behaviour, and thus their executive or control functions represent an 'inner' teacher. Vygotsky (1978) suggested that all such higher-order functions originate as actual relationships between individuals, thus before students can 'internalize' these skills they must develop them explicitly with others. Working collaboratively in small groups requires students to elaborate and verbalise their thinking. Discussion and questioning within a supportive group leads students to construct a 'scaffolding' framework for each other, which enables them to solve problems collaboratively before they can solve such problems individually (Forman and Cazdan, 1985; Wheatley, 1991). The type of questions asked by the teacher, which emphasise the need for self-evaluation, are internalised by the students and used to guide and monitor their progress (Schoenfeld, 1987; Gray, 1991).

The nature and importance of self-evaluation, reflecting back not only on the solution but also on the thought processes which led to it, suggests that teaching students to assess themselves is an integral part of teaching modelling.

2 PROJECT AND METHODOLOGY

The Use and Practical Applications of Mathematics Project was funded by the UK Welsh Office during 1991/92. The project aimed to develop approaches and materials to teach and assess the thinking skills involved in using and applying mathematics in practical modelling situations.

A network of eight schools was established to conduct the research. The action research paradigm was chosen owing to the novel approaches involved in many of the activities, which would require teachers to develop new teaching styles as the project progressed.

Two teachers from each school attended an intensive course on practical mathematics at the start of the project. They then returned to school to assist their colleagues in teaching experimental practical lessons, which were monitored using a self-evaluation form. University researchers acted as participant observers in many of the lessons. The teacher-researchers attended regular network meetings to discuss progress, and to develop approaches and resources.

3 THE MODELLING ACTIVITIES

The tasks were designed to be rich practical activities which would allow students to pose their own problems from real-world situations, and formulate their own mathematical models. A variety of mathematical approaches were possible, leading to predictions which could be tested in the real world. Tasks extended over several lessons, and a wide choice of practical equipment was available for students to use. The aim of the teaching was to develop the thinking skills associated with modelling.

The tasks were produced in a variety of formats, with differing degrees of structure – to correspond to the background and ability of the students concerned and the philosophy of mathematics held by the teacher leading the lesson. Two typical tasks are listed below (for further examples see Tanner & Jones, 1993).

The ice cream problem

You are in charge of the ice cream stall at the summer fête. One 2 litre tub of ice cream costs £2.29. One packet of 20 cones costs 89p. Each cornet takes one cone and one scoop of ice cream. How much should you charge for each ice cream cornet in order to make the most profit?

Do you need any further information? How will you obtain this?

Pipes and Wires

Electricity cables are often carried around buildings inside a special casing called conduit or ducting. What shapes could these casings be?

Two students decide to model this problem using straws (to represent the wires) and strips of thin card (to represent the casings). Investigate possible sizes and shapes of casing.

The tasks were attempted by whole classes of students, from across the ability range, in each year group between the ages of 11 and 16.

Initially a wide range of approaches were employed, which provided differing degrees of structure for the students. A degree of consensus developed gradually, namely that teachers should provide 'scaffolding' in the early stages to enable students successfully to complete tasks which would otherwise be beyond them (see Bruner, 1985). It was considered that such 'scaffolding' was best provided through questioning and discussion. However, the form of such questioning and discussion varied greatly from teacher to teacher. Some teachers tended to ask questions which were specific to the activity, which often left the students responsible for abstracting those elements which were generalisable. Others asked general questions in a systematic way and encouraged students to transform such external processes into internal processes which could be used for self regulation and planning through inner speech (see Wertsch and Stone, 1985). In the second case, students were generally more able to assimilate such processes into their conceptual schema, and build on them in later problems, thus allowing the external scaffolding to be reduced.

4 DEVELOPING ASSESSMENT FRAMEWORKS FOR THE ACTIVITIES

For each activity developed during the project, the teaching approaches employed were recorded and analysed, together with the students' responses. Informal interviews with students and teachers were recorded after the lessons. Students were observed and interviewed whilst working on the activities, in order to determine their perceptions of the situations and the tactical and executive strategies which they were employing to obtain a solution. This information was then used to identify clearly-definable levels of performance specific to each practical task.

Task-specific criteria derived from observed behaviours were written under the headings Strategy, Communication, and Logic, using a ten-point scale. Task-specific behaviours which had been commonly observed in middle-set 14 year old students were equated to level 6, in parallel with the national assessment structure in use in England and Wales (see DES/WO, 1991). Other observed levels of performance were then placed on the scale using this point as a benchmark. The resulting behavioural criteria for performance under the three headings formed a criterion-referenced assessment framework specific to each task.

Transcripts and task criteria from all the tasks (38 in all) were then analysed to develop general criteria for practical problem solving. These general criteria were used to guide assessment when an idiosyncratic problem solver pursued an unusual line which did not fit the specific behavioural criteria. The general criteria were also of use when devising assessment frameworks for new tasks. Examples of assessment frameworks appear in the Appendix.

5 ASSESSING MODELLING ABILITY

The frameworks provided a basis for assessing students' modelling abilities. In order to gain a valid assessment, a variety of modes of assessment were used. Informal, unrecorded assessments were made by the teacher within the normal flow of a lesson. Oral assessment occurred during informal discussions and the 'reporting back' sessions. Student performance was matched against the task-specific criteria in the framework. The frameworks were also written in 'student-friendly' language for use during self-assessment.

When the students' written accounts of their work were examined by groups of teachers, the assessment frameworks proved to be easy to use and were reliable, in that results were generally consistent between assessors. Assessment of written work alone, however, was considered to lack validity, partly due to the lack of consistency in presentation between teachers. The degree of structure provided to students, by the teacher, at the start or during the task varied considerably, particularly in the initial stages of the project. In some lessons so much structure had been provided by the teacher that it was not possible for students to demonstrate ability in the strategy strand – all the thinking in this area had been done by the teacher. Other teachers, wishing to make an 'accurate' summative assessment, were unwilling initially to make any kind of intervention. This often left inexperienced students floundering, unsure of what was expected of them. The degree of structure provided by the teacher was not apparent in individual scripts, and it was necessary to use a variety of assessment techniques, depending on the particular context, to ensure validity.

6 SOCIALISATION INTO MODELLING

Different groups of students reacted very differently to each other when presented with practical situations to model. Even groups which had experience of 'pure' investigative work and 'open-ended tasks' were unsure how to proceed, and often failed to connect such work with practical situations.

Supposedly 'open' situations are not as open and free as they seem. To learn mathematics is not simply to learn a list of skills and routines. To learn mathematics is to take part in a process of *socialisation* or acculturation (Schoenfeld, 1987). Our students need to learn ways of acting and thinking which are acceptable in the society of mathematicians, not just the society of the traditional mathematics classroom. Unfortunately many students associate acceptable ways of acting and thinking mathematically with dependence and repetition (Lester and Kroll, 1990; Gray, 1991). Even when independence, initiative and creativity are seen as desirable characteristics, students often remain unaware of the nature of mathematically-acceptable solutions to modelling exercises. For many teachers and students the introduction of modelling and problem solving involves a reappraisal of their philosophy of mathematics – mathematics is not a body of knowledge to be memorised and reproduced, but rather a process of enquiry and generalisation.

It is sometimes argued that it is not possible to assess open-ended work due to the variety of outcomes which might be found within a class. This implies that a good solution would be impossible to recognise when presented. This is not the case. Mathematicians work to a set of assumptions, often unstated, relating to generality, economy and elegance. Teaching students mathematics must include socialising them into these assumptions. To be able to operate in an open, practical or modelling situation, students must be aware of the general nature of qualities to be found in a desirable solution, and be able to assess their own work against these criteria.

Assessment is a significant factor in the socialisation of students into the world of mathematicians, and it is through self-assessment that students learn to model practical situations. In order to use executive strategies in the course of modelling situations, students must first be able to monitor and evaluate their own work.

7 SOCIALISING STUDENTS THROUGH SELF-ASSESSMENT

Students were encouraged to assess their own work against the students' self-assessment framework for the activity. This formed the basis for a dialogue between the student(s) and the teacher, to validate any informal assessments made and to help students understand the criteria against which they were being assessed. Some teachers used the self-assessment frameworks before the write-ups were completed, so that the students had an opportunity to amend and improve their work.

The accuracy of the self-assessments made by students was variable from class to class. Self-assessment requires both an understanding of the nature of 'good solutions' and a degree of reflection on process which was clearly absent from some classrooms. It was apparent that some teachers had been much more successful than others in transmitting 'modelling culture'. Students who were unskilled in reflecting on their work, and engaging in self-assessment, tended to report themselves at a higher level than their teachers considered appropriate. The dialogue between teacher and student which followed such over-reporting made a useful contribution to the socialisation process. Students reported finding the frameworks helpful in showing them what was important, both in carrying out an investigation and in writing it up, and that following the use of a framework they had a clearer understanding of how their work could be improved in future.

Students were more successful in self-assessment if they had been involved in the process of peer assessment. In this process students were encouraged to present their interim findings to the class. Ground rules were carefully established – the purpose of 'reporting back' was to improve each group's work through discussion and constructive criticism. Students could ask for explanation of points they did not understand – 'How did you know...? Why did you do..? Will that always be true...?' – but any criticism had to be justified with reasons and, preferably, accompanied by suggestions for improvement. The value that was placed on this 'reporting back' was shown by the competition that came to exist between the students to have their work selected for discussion. Initially, teachers led the questioning during the presentation of work but, with encouragement, the students took a major part, incorporating the type of questions asked by the teacher with their own. Reporting back provided an excellent opportunity to monitor the progress made by groups, and to disseminate good ideas and strategies across the class. More importantly, it provided a mechanism for the explicit rehearsal of metacognitive skills between individuals.

Following completion of the task students were required to write up their work individually, using guidelines which had been discussed with the class. Selected groups also presented their final report to the class. A vital question which had to be answered was 'If I were to do this investigation again what would I do differently?' This encouraged 'looking back' and allowed consideration of the elegance of different solutions, economy of approach to a task, and so on. Assessing the work of others helped students to identify the nature of a good solution, and reporting back made them concentrate on what was important in their own work. By learning how to assess the work of others, students learned to reflect on and assess their own.

8 INTERFERENCE BY OBSERVERS

No attempt was made by the observers to be 'invisible' in the classroom, rather we regarded ourselves as full participants in the experience and interfered significantly with the processes we were observing. By asking students to describe the strategies they were using, and how they had selected those strategies, we were asking them to monitor and evaluate their work in a manner which they might not otherwise have done. By asking such questions we gave them importance, and placed emphasis on planning, monitoring and evaluating. Teachers often asked similar questions themselves. The consensus was that such questions were a part of good practice when teaching modelling, as students gradually internalised the mode of questioning used in such exchanges (Forman and Cazden, 1985).

9 CONCLUSION

Mathematical-modelling skills are most successfully acquired when 'scaffolding' is provided which structures the task into steps that the student can just manage. The questions and discussions which form this scaffolding should be general rather than task-specific, to facilitate their internalisation into students' schema.

Discussion of modelling strategies between the teacher and students, and between the students themselves, helps to clarify the decisions made at each stage, and the reasons for them. Such discussions make the students more aware of their own thinking, and help to develop the metacognitive skills of inner speech.

Many students (and teachers) need to be socialised into modelling culture. When students approach a task in the classroom, many of the decisions about processes and outcomes have already been made. They have been determined by the common assumptions of the classroom environment, and assessment plays a significant role in establishing these. Assessment should be an integral part of the teaching process. It should inform teaching, by giving the teacher a clearer perspective of the student's stage of development, but it should also provide a channel of communication between teacher and student, through which socialisation into modelling culture can occur.

Reporting back to the class through the process of peer assessment in an atmosphere of mutual support helped students to become aware of what was important in their own work. This is an important aspect of their socialisation as mathematicians. Reflecting on the work of others led students inevitably to reflect back on their own work. They seemed able to use the work of others as a mirror to examine their own thought processes. Through reflecting on and assessing the work of others, students learned to evaluate and regulate their own thinking.

The metacognitive skills of planning, monitoring and evaluating are integral to successful modelling. Self and peer assessment provided a mechanism for the explicit rehearsal of skills which, through practice, became a routine part of the student's repertoire. Self-assessment is at the heart of the modelling process, and learning the skills of self-assessment helps students to learn to model.

REFERENCES

Alston, A S and Maher, C A (1987). Children's connections among representations of mathematical ideas. In Bergeron, J C, Herscovics, N and Kieran, C (ed), *Proceedings of the Eleventh Annual Meeting of the International Group for the Psychology of Mathematics Education.* Vol 2, 257-266.

André, T (1989). Problem solving and education. In Moon, B and Murphy, P (ed), *Developments in Learning and Assessment.* London: Hodder & Stoughton/OU, 60-74.

Brown, A (1987). Metacognition and other mechanisms. In Weinert, F E and Kluwe, R H (ed), *Metacognition, Motivation and Understanding.* Hillsdale: Lawrence Erlbaum, 65-116.

Bruner, J S (1985). Vygotsky: a historical and conceptual perspective. In Wertsch, J V (ed), *Culture, Communication & Cognition: Vygotskian Perspectives.* Cambridge: Cambridge University Press, 21-34.

Campione, J C (1987). Metacognitive components of metacognitive research with problem learners. In Weinert, F E and Kluwe, R H (ed), *Metacognition, Motivation and Understanding*. Hillsdale: Lawrence Erlbaum, 117-140.

DES\WO (1991). Mathematics in the National Curriculum. London, HMSO.

Dreyfus, T (1990). Advanced mathematical thinking. In Nesher, P and Kilpatrick, J (ed), *Mathematics and Cognition: A Research Synthesis.* Cambridge: Cambridge University Press.

Forman, E A and Cazden, C B (1985). Exploring Vygotskian perspectives in education: the cognitive value of peer interaction. In Wertsch, J V (ed), *Culture, Communication & Cognition: Vygotskian Perspectives.* Cambridge: Cambridge University Press, 323-347.

Gray, S S (1991). Ideas in practice: metacognition and mathematical problem solving. In *Journal of Developmental Education,* **14**, 24-28.

Hirabayashi, I and Shigematsu, K (1987). Metacognition: the role of the "inner teacher". In Bergeron, J C, Herscovics, N and Kieran, C (ed), *Proceedings of the Eleventh Annual Meeting of the International Group for the Psychology of Mathematics Education.* Vol 2, 243-249.

Lester, F K (1987). Why is problem solving such a problem? In Bergeron, J C, Herscovics, N and Kieran, C (ed), *Proceedings of the Eleventh Annual Meeting of the International Group for the Psychology of Mathematics Education*. Vol 2, 257-266.

Lester, F K and Kroll, D L (1990). Teaching students to be reflective: a study of two grade seven classes. In Booker, G, Cobb, P. and de Mendicuti, T N (ed), *Proceedings of the Fourteenth Annual Meeting of the International Group for the Psychology of Mathematics Education.* Vol 1, 151-158.

Manning, B H (1984). A self-communication structure for learning mathematics. In *School Science and Mathematics*, **84**, 43-51.

Schoenfeld, A H (1985). *Mathematical Problem Solving*. Orlando: Academic Press.

Schoenfeld, A H (1987). What's all this fuss about metacognition? In Schoenfeld, A H (ed), *Cognitive Science and Mathematics Education*. Hillsdale: Lawrence Erlbaum,189-215.

Silver, E A (1987). Foundations of cognitive theory and research for mathematics problem-solving instruction. In Schoenfeld, A H (ed), *Cognitive Science and Mathematics Education.* Hillsdale: Lawrence Erlbaum, 33-61.

Tanner, H and Jones, S (1993). *Hands on Maths.* Swansea: University College of Swansea Press.

Vygotsky, L S (1978). *Mind in Society: The Development of Higher Psychological Processes.* Cambridge: Havard University Press.

Wertsch, J V and Stone A C (1985). The concept of internalization in Vygotsky's account of the genesis of higher mental functions. In Wertsch, J V (ed), *Culture, Communication & Cognition: Vygotskian Perspectives.* Cambridge: Cambridge University Press, 162-179.

Wheatley, G H (1991). Constructivist perspectives on science and mathematics learning. In *Science Education* **75,** 9-21.

APPENDIX 1 – *ICE CREAM - Assessment Framework AT1*

Level	Strategy	Communication	Logic
2	Follows instructions carefully, finds the number of cornets that can be made from one tub of ice.	Responds to questions and describes what s/he is doing sufficiently well for the teacher to understand.	Asks and responds to questions such as 'What would happen if the tub was larger?'
3	Identifies a problem associated with measurement and suggests how to solve it, eg: the scoop should be levelled off to hold the same volume each time.	Expresses the question clearly in their own words. Describes the work being done and records the results in such a manner that they can be retrieved later.	Try out some new examples having changed the situation, eg: changed the size of the scoop. Check the results and consider whether they seem sensible.
4	Suggests a method of measuring the number of scoops in the tub.	Describes the model being used in mathematical language. Uses clear diagrams. Records findings systematically.	Makes simple predictions, eg: for the same scoop and tub you will always make the same number of ices.
5	Attempts to price the cones at a profit. Considers the accuracy of the result.	Uses simple mathematical techniques to display and to help interpret findings.	Makes and tests simple generalisations, eg: if twice as many people attend then we will sell twice as many ice creams.
6	Identifies some of the variables affecting the number of ice creams that might be sold, eg: buyer's age, attendance, weather; and investigates one of these by means of a survey or otherwise.	Describes the predictions made linking two variables using every-day language, eg: most of the young people attending will buy one ice cream.	Justifies the generalisation in some way, eg: by referring to the results of the survey.
7	Uses the data obtained to predict the probable level of sales and the profit which would be made.	Interprets results with some degree of precision, eg if the attendance is doubled then the sales of ice cream will almost double..	Tests hypotheses by further experiment and comments on their validity, eg: compares predictions with records of sales in previous fayres or with commercial data.
8	Investigates the effect of changing other variables.	Describes relationships in mathematical terms, eg the number of ices sold is inversely proportional to the temperature.	Evaluates the model being used. Uses the results to suggest improvements and further questions for exploration. Tests any relationships found by searching for counter-examples.
9	Analyzes the data obtained from the investigation.	Draws conclusions and forecasts the probable sales of ice cream.	Quantifies any relationships found.
10	Extends the task, eg investigates the forecasting methods used commercially to set prices/profit margins.	Explains and attempts to justify the conclusions reached.	Attempts to justify or prove the experimental formula.

APPENDIX 2 – ***Pipes and Wires - Assessment Framework AT1***

Level	Strategy	Communication	Logic
2	Follows instructions carefully and finds the number of wires inside a casing.	Responds to questions and describes what s/he is doing sufficiently well for the teacher to understand.	Asks and responds to questions such as 'What would happen if the casing was longer?'
3	Identifies a problem associated with measurent and suggests how to solve it, eg: it is easier to measure the length of the casing before folding it up.	Expresses the question clearly in their own words. Describes the work being done and records the results in such a manner that they can be retrieved later.	Try out some new examples having changed the situation, eg: altered the casing. Check the results and consider whether they seem sensible.
4	Collects and records at least three items of data, eg: finds the number of wires in three different casings.	Describes the model being used in mathematical language. Uses clear diagrams. Records findings systematically.	Makes simple predictions, eg: if the perimeter stays the same then you will always get the same number of wires in.
5	Applies method to find the number of wires in a variety of different casings.	Uses simple mathematical techniques to display and to help interpret findings.	Makes and tests simple generalisations, eg: if we double the perimeter then it will hold twice the number of straws.
6	Identifies some of the variables affecting the number of wires contained and investigates one whilst holding the others fixed.	Describes the predictions made linking two variables using every-day language, eg: increasing the perimeter increases the number of wires inside.	Justifies the generalisation in some way, eg: by referring to the results.
7	Draws a sensible line on a graph connecting, for example, the number of wires and the perimeter of the casing.	Interprets results with some degree of precision, eg: if the perimeter of the casing is doubled then the number of wires goes up by more than double.	Tests hypotheses by further experiment and comments on their validity, eg: extrapolating from a graph and checking the results.
8	Having explored the effect of changing one variable, selects another to investigate whilst holding all the others fixed.	Describes relationships in mathematical terms, eg: the area of the casing is proportional to the number of wires.	Evaluates the model being used. Uses the results to suggest improvements and further questions for exploration. Tests any relationships found by searching for counter-examples.
9	Derives a formula	Relating two of the variables.	Quantifies an algebraic relationship from the results, eg: $N = 0.7 \times A$
10	Extends the task, eg investigates the relationship between length, area and volume for solids.	Explains and attempts to justify the conclusions reached.	Attempts to justify or prove the experimental formula.

22

Future of mathematical modelling in the classroom

Solomon Garfunkel
Consortium for Mathematics and Its Applications, Lexington, USA

SUMMARY

This paper – looking into the future – deals with the question of how to introduce and develop the necessary mathematical concepts in a natural and meaningful way under the presumption of a given technology-enhanced modelling and application based curriculum. It describes modelling problems already developed, which enable the students to develop their thinking skills, intellectual curiosity, and algorithmic skills as part of the process of learning mathematics. Furthermore, a framework for a high school curriculum is proposed, which emphasises a mathematics curriculum organised around crucial themes and concept structures, and names applications to all concept structures.

1 CURRICULUM PLANNING

When I work with colleagues on curriculum projects, there is a question I like to ask in our planning sessions. What if we win! This sounds a bit like a variant of the Chinese curse – may all your dreams come true – and it is. While it is certainly impossible (at least for me) to predict accurately the time frame (2 years, 5 years, 10 years...), I honestly believe that we will eventually win.

What do I mean by that? Basically, two things. First, we will have at our disposal all of the computational power, interactivity and telecommunication facilities we can reasonably use. Second, that applications and modelling will become the core organising principle of the primary and secondary mathematics curricula. Since this is my own view of the future, I will not spend a great deal of time arguing why I believe this will come to pass. However, looking at the technological developments around us makes the first point almost certain. The Dutch experience, the work of de Lange and his colleagues at the Freudenthal Institute, and the positive attention and publicity it has received, yields strong evidence for the second point.

Make no mistake, while the future I envisage is certainly a victory for those of us who have laboured long and hard in these directions, it is a mixed blessing. On the one hand it permits the implementation of pedagogical and assessment changes around which we have reached some consensus – hands-on, activity-based, cooperative learning, and suchlike. On the other hand it opens up a Pandora's box – how do we develop the necessary mathematical themes and (when and if) do we differentiate among students who intend to go on in mathematics and science and those who do not?

Actually, the future availability of high-speed computation, graphics and communication brings its own Pandora's box. The good news is clear. Modelling curricula are frequently criticised for being simplistic. The real world is more complicated and has more data points, while our classroom applications trivialise serious problems. These criticisms will certainly disappear. Given the computer power at our disposal, it is only a question of time and some cleverness on our part in using these tools to attack larger problems.

Communications also open up possibilities. It is easy to imagine examples in the spirit of KIDNET, collecting acid rain data simultaneously from thousands of students all over the world. In addition, interactivity can bring the visual power of video directly to the students.

The Pandora's box, of course, is symbol manipulation. We have not, as yet, reached consensus on the weight to give paper-and-pencil algorithms, which in practice will be done by calculator and/or computer. I suspect, however, that this problem, as most technological issues, will *de facto* sort itself out. We will become increasingly dependent upon the technology, and introduce algorithms only for developmental and conceptual purposes. However, this brings us back full circle to the hard part. Given a technologically-enhanced modelling and application based curriculum, how do we introduce and develop the necessary mathematical concepts in a natural an meaningful way so that students can make them their own?

THE ARISE PROJECT

In a recent COMAP proposal to the National Science Foundation, *Applications/ Reforms in Secondary Education (ARISE)*, to produce an integrated curriculum for year 9-11 organised by modelling themes, we contrasted the old with the new.

Mathematics is widely perceived to be concerned with the solution of the kinds of problems shown below, because this is the material that is taught and this is the material that is tested on mathematics 'skills and proficiency' tests.

- Simplify: $\frac{1}{x}+\frac{1}{x^2}-1$
- Solve for x: $3x-(2x+5)=-x+15$
- How old is Mary if Jane is 3 years older than Mary was 5 years ago, and the sum of their ages is 9?

No wonder mathematics is misperceived! This sterile environment, stripped of a framework of important mathematical themes, concerns, and applications, is doomed to failure. Teaching an essentially unchanging high-school mathematics curriculum for over 50 years has successfully alienated nearly all but the 'mathematically inclined' who, by contrast, and in defiance of the perception of the majority, seem to thrive on mathematics.

The design for a new high-school curriculum is predicated on the assumption that what is required is not merely a collection of proficiencies and skills, but exposure to mathematics' rich conceptual structure, and perhaps most importantly, its applications. We recognise that it may be harder to achieve these more ambitious goals than mindlessly to encourage proficiency on a well-defined collection of mechanical skills that can be mastered algorithmically. However, the potential payoff is more than worth the effort.

To show the contrast between what could be done and what is done, consider the following problems.

- An estate left to three children consists of a boat, a summer home, and a collection of rare books. Discuss methods of dividing the estate fairly (Garfunkel *et al*, 1990).

- In ranking a collection of high-school basketball teams, sportswriters produced the results below (Malkevitch, 1989).

	A1	A2	A3	A4	A5	A5
	A4	A5	A2	A2	A2	A3
	A5	A4	A5	A5	A4	A4
	A3	A3	A4	A3	A3	A2
	A2	A1	A1	A1	A1	A1
VOTES	18	12	10	9	4	2

  Devise a method for deciding which team deserves to be ranked highest.

- A small corporation has four stockholders – A, B, C and D – who control 18, 22, 34, and 26 percent of the stock respectively, each of whom casts a vote weighted by the percentage of stock they control. Important decisions are taken on the basis of at least a majority of the weight cast. Is there a way of representing the 'power' of the stockholders in the decision making process? If A and D enter into a secret agreement always to vote together, how are the power relations affected? (Garfunkel *et al*, 1990).

- Shown below are several five-digit Zip codes and the machine-readable characters that represent them. Work with the people in your group to try to crack the code. What symbols, for example, represent 5 or 7? Try to determine the symbols that represent each of the ten digits of our number system (Malkevitch *et al*, 1991).

17011

20078

80328

19006

05445

- The voting rule for the United Nations Security Council is as follows.

  - Any of the five permanent members can cast a veto.

  - Nine out of the fifteen members must vote in favour of an issue in order for it to pass.

  a Describe the minimal winning coalitions for this voting system.

  b Desribe this system as a weighted voting system.

  c Compute the Banzaf power index for this system.
  [Hint: One does not need to examine each of the $2^{15} = 32\ 768$ combinations of Y and N votes for pivots. There are only two types of voters, permanent members and temporary members. One can thus use the counting techniques from an algebra, a probability, or a discrete mathematics course (Lucas, 1992).]

- The surgery department of the Healthy-Life Hospital categorises operations as 'minor' and 'major' depending on whether they have a duration of 45 minutes or $1\frac{1}{2}$ hours. The hospital charges $125 for a minor surgery and $175 for a major surgery. If the hospital has six operating rooms, in use on the average for six hours per day, advise the management how to schedule the number of surgeries per day so as to maximise the revenue (Sriskandarajah, 1992).

3 CURRICULUM DESIGN

The understanding and solution of such problems requires the same skills we currently teach (usually in a vacuum). Yet these problems enable students to develop their thinking skills, intellectual curiosity, social skills, and algorithmic skills as part of the process of learning mathematics. Although symbolic and geometric representations are part of the strategy of solving these problems, they are not ends unto themselves, as they might be in a totally mathematically driven environment. However, the high-school mathematics environment exists for much broader reasons than the mathematics pipeline – the training of future mathematicians, scientists and engineers. The curriculum which will be developed is rooted in this belief, and samples of what one can carry out when the skills, concept, historical, and applications portions of mathematics are integrated are illustrated in the table below. This table is not a curriculum design.

Rather it represents the starting point for our work – a framework which emphasises a mathematics curriculum organised around crucial themes and concept structures.

Concept structure	Mathematical skills	Applications
Fairness (treated the first year)	1 Solving linear equations 2 Inequalities 3 Matrices 4 Review of decimal and fraction arithmetic 5 Functions 6 Graphing functions 7 Probability 8 Statistics 9 Graph theory 10 Modelling	1 Fair games (a) Lotteries (b) Sociology Psychology 2 Fair elections (a) Political elections (b) Ranking economic alternatives 3 Fair division (a) Continuous objects (b) Discrete objects (i) Estates (ii) Divorce settlements 4 Apportionment (a) House of Representatives (b) European Parliaments 5 Sports
Codes (treated the first year)	1 Number in bases other than 10 (binary) 2 Modular arithmetic 3 Distance 4 Matrices 5 Counting techniques 6 Modelling 7 Solving linear equations	1 Zip code 2 Universal product code 3 Pictures from outer-planets 4 Compact discs 5 High definition television 6 Modems 7 ISBN numbers 8 Barcodes 9 Cryptography
Concepts of space (treated in all three years; emphasised in 2nd year)	1 Euclidean geometry 2 Parallel postulate 3 Non-Euclidean geometry 4 Rôle of proofs 5 Relation of mathematics to science	1 Physics 2 Perception 3 Psychology

Concept structure	Mathematical skills	Applications
Symmetry, pattern, and asymmetry (treated in all three years)	1 Distance 2 Geometric transformations 3 Tilings 4 Lattice points 5 Graph theory 6 Functions 7 Fractals 8 Angles 9 Polygons 10 Trigonometry 11 Area 12 Volume	1 Analysis of fabrics, design, wallpapers 2 Crystals 3 Architecture 4 Dating techniques 5 Physics 6 Cartography
Optimisation (treated in all three years)	1 Functions 2 Inequalities 3 Systems of equations 4 Matrices 5 Graph theory 6 Counting 7 Algebraic simplification	1 Break-even analysis 2 Linear programming 3 Shortest paths 4 Scheduling of processors and people 5 Travelling salesman problem 6 Chinese postman problem
Mathematical modelling (treated in all three years)	1 Deterministic *v* stochastic phenomena 2 Continuous *v* discrete phenomena 3 Physical models 4 Functions 5 Graphs 6 Matrices 7 Differences between mathematics and science 8 Rôle of proof in mathematics 9 Empirically based models	1 Social sciences 2 Sciences 3 Humanities 4 Engineering 5 Computer science 6 Business 7 Leisure activities

Concept structure	Mathematical skills	Applications
Change and growth (treated in the third year)	1 Solving equations 2 Functions 3 Graphing 4 Recursion 5 Pattern recognition 6 Difference equation 7 Algebraic simplification 8 Laws of exponents 9 Geometric progression 10 Arithmetic progression	1 Charge and payment 2 Sinking funds 3 Retirement and college tuition funds 4 Mortgages 5 Drug uptake 6 Animal harvesting
Risk (treated in the second year)	1 Probability 2 Data analysis 3 Exponential functions 4 Inequalities	1 Insurance 2 Sports 3 Health 4 Ecology 5 Technology
Conflict and competition (treated in thc third year)	1 Equations 2 Graphing 3 Graph theory 4 Packing and covering 5 Probability	1 Food webs 2 Resource management 3 Psychology and sociology 4 Political science 5 Voting procedures 6 Electoral college 7 Business negotiations 8 Scheduling theory

4 FOR ALL – NOT FOR SOME

Two things should be re-emphasised here. This is intended as a core curriculum for grades 9-11 – a curriculum for all students. Any differentiation based on career goals and college plans will take place in grade 12. Moreover, we are serious about the thematic nature of the curriculum. The mathematical techniques will be introduced as needed to serve the concept structure. Implicit in this 'article of faith' is our conviction that techniques for which we cannot find a legitimate use will *not* be introduced for their own sake.

This naturally puts a strong burden on a subsequent coursework. For those students going on to study mathematics at the college level, it is important to create an honest and appropriate 12th-year course, which can serve realistically as a bridge to calculus. We feel it is important in this year to reveal and organise this course around the underlying mathematical structure – while consistently emphasising the applicability of the subject matter.

Let me re-emphasise that the hard work is yet to be done. It is one thing to talk about building mathematical skills in this context, but it is another to do it. It is one thing to talk about effective and appropriate use of technology, but it is another to carry it out. That is, however, our task. We are committed to this organisation for a core 9-11 programme, with differentiation for students in year 12. We share with our international colleagues the opportunity and responsibilities of making this approach work for students – and the enormous job of preparing teachers to work in this new educational environment. However, this is the future as I see it. Since I began with a Chinese curse, I will end with one – We live in interesting times.

REFERENCES

Garfunkel, S *et al* (1990). *For All Practical Purposes: Introduction to Contemporary Mathematics,* 2nd ed. New York: Freeman.

Lucas, W (1992). *Fair Voting: Weighted Votes for Unequal Constituencis. HistoMAP Module 19.* Lexington: COMAP.

Malkevitch, J (1989). *The Mathematical Theory of Elections. HiMAP Module 1.* Arlington: COMAP.

Malkevitch, J, Froelich, G and Froelich, D (1991). *Codes Galore, HistoMAP Module 18.* Arlington: COMAP.

Sriskandarajah, S (1992) *Optimality Pays: An Introduction to Linear Programming. HistoMAP Module 20.* Lexington: COMAP.